Cambridge Elements ≡

Elements in the Philosophy of Science
edited by
Jacob Stegenga
University of Cambridge

SCIENTIFIC PROGRESS

Darrell P. Rowbottom
Lingnan University

Shaftesbury Road, Cambridge CB2 8EA, United Kingdom

One Liberty Plaza, 20th Floor, New York, NY 10006, USA

477 Williamstown Road, Port Melbourne, VIC 3207, Australia

314–321, 3rd Floor, Plot 3, Splendor Forum, Jasola District Centre,
New Delhi – 110025, India

103 Penang Road, #05–06/07, Visioncrest Commercial, Singapore 238467

Cambridge University Press is part of Cambridge University Press & Assessment,
a department of the University of Cambridge.

We share the University's mission to contribute to society through the pursuit of
education, learning and research at the highest international levels of excellence.

www.cambridge.org
Information on this title: www.cambridge.org/9781108714433

DOI: 10.1017/9781108625753

First published 2023

A catalogue record for this publication is available from the British Library.

ISBN 978-1-108-71443-3 Paperback
ISSN 2517-7273 (online)
ISSN 2517-7265 (print)

Cambridge University Press & Assessment has no responsibility for the persistence
or accuracy of URLs for external or third-party internet websites referred to in this
publication and does not guarantee that any content on such websites is, or will
remain, accurate or appropriate.

Scientific Progress

Elements in the Philosophy of Science

DOI: 10.1017/9781108625753
First published online: September 2023

Darrell P. Rowbottom
Lingnan University

Author for correspondence: Darrell P. Rowbottom,
darrellrowbottom@ln.edu.hk

Abstract: What constitutes cognitive scientific progress? This Element begins with an extensive survey of the contemporary debate on how to answer this question. It provides a blow-by-blow critical summary of the key literature on the issue over the past fifteen years, covering the central positions and arguments therein. It also draws upon older literature, where appropriate, to inform the treatment.

The Element then enters novel territory by considering meta-normative issues concerning scientific progress. It focuses on how the standards involved in assessing progress arise. Does science have aims, which determine what counts as progress, as many authors assume? If so, what is it to be an aim of science? And how does one identify such things? If not, how do normative standards arise? After arguing that science does not have overarching aims, the Element proposes that the standards are ultimately subjective.

Keywords: scientific progress, scientific realism, meta-normativity, aim of science, scientific change

ISBNs: 9781108714433 (PB), 9781108625753 (OC)
ISSNs: 2517-7273 (online), 2517-7265 (print)

Contents

1 The Contemporary Debate on Scientific Progress: What Constitutes Cognitive Progress?

> Although it is nearly uncontroversial that science makes progress of some sort or other, it is far from uncontroversial what scientific progress consists in.
>
> (Dellsén 2016: 72)

Prima facie, pre-philosophically, some developments in science have constituted intellectual advances, whereas others have not. The detection of gravitational waves, reported by Abbott et al. (2016), is a plausible example of an advance; most think that this was a 'step forward' for science qua an intellectual endeavour. The hundreds of papers on polywater in the 1970s seem instead to have been a 'step backwards' in the same dimension. Scientists now think that what was taken to be a polymer, composed of H_2O monomer units, was really something else.[1]

Some developments also appear to have been more intellectually significant than others. For instance, the discovery that polywater does not exist seems less important than the advent of Bohr's (1913) model of the atom, which gave a potential explanation of the absorption/emission spectra of hydrogen and provided a means to predict unknown spectral lines of several other elements, inter alia.[2] Thus it is natural to think that we could rank scientific changes in terms of how much of an improvement they constituted.

We might therefore adopt the following approach to exploring what kinds of changes are responsible for *cognitive progress* in science, as Laudan (1977) called it. We might trust that we can reliably identify instances when progress has occurred and that we can reliably rank them in order of significance. We might then generate lists of progressive episodes and orderings of the episodes with respect to the degree of improvement they constituted. Having done this, we might perform comparisons to determine what constitutes scientific progress. We might ask questions such as 'What are the similarities between the cases where cognitive progress occurred?' and 'What differentiates cases where substantial progress was made from situations where minimal progress was made?'

It is also natural to use thought experiments and counterfactual judgements to assist in the task. We might imagine what scientists could have done differently in a historical scenario. Or we might conceive of a hypothetical situation where

---

[1] Sweat contamination is commonly thought to be to blame, following Rousseau (1971). However, van Brakel (1993) argues that a reaction between water and silica was responsible.

[2] Key was Bohr's reduction of the Rydberg constant to an expression involving electron mass, electron charge, and several physical constants. For more on the background to and the significance of this episode, see Rowbottom (2019: ch. 4), Aaserud and Heilbron (2013), and Heilbron and Kuhn (1969).

scientists face a dilemma and consider which choice would result in more progress.

This is how the contemporary debate about cognitive scientific progress usually proceeds; evidence to this effect will become apparent. Questions such as 'Does science really cognitively progress?' and 'What is the source of the standards for judging what is cognitively progressive?' are only touched upon in modern philosophy of science. I believe that such meta-normative questions are pressing, for reasons that will emerge. Sections 2 and 3 of this Element will focus on them, in the hope of reorienting the ongoing discussion.

For the moment, I will say something in defence of the standard approach. All inquiries need starting points. One cannot have a view from nowhere, any more than one can have a God's-eye view. Furthermore, it is natural to begin with the assumption that something is present when it seems to be, and to see if you can learn more about it. Philosophers have done this with putative things as disparate as properties, time, knowledge, minds, and selves. And if philosophy cognitively progresses, one of the ways it does so is by charting the possibilities about such things.

In this section of the Element, therefore, I will cover the status quo on cognitive scientific progress. To prepare the stage for doing so, I will next do two things. First, I will explain and justify the precise scope of this survey. Second, I will present several key distinctions that will be useful throughout.

1.1 Survey Scope

This survey is limited in two respects. First, it concerns only literature on cognitive progress. You might wonder why this is. The short answer is that philosophers of science have focused on this kind of progress because it seems to be a distinctive feature of science. This is not to deny that other sorts of progress are possible in science or that they are worthy of philosophical study. For instance, science might progress or regress in a moral sense. But work on moral issues in science is typically labelled as 'ethics' or 'science studies' as a matter of convention. In the philosophical community, 'scientific progress' is customarily taken only to refer to a cognitive issue. To see this, consult the entry on 'scientific progress' in the *Stanford Encyclopedia of Philosophy* (Niiniluoto 2019) or the most recent article on the topic in *Philosophy Compass* (Dellsén 2018a).[3]

---

[3] I recommend both as further reading. Dellsén (2018a) gives a concise overview of the dominant theories of progress, and the main considerations for and against each; it is an excellent starting point, which gets to the core of the issues while minimising complications. Niiniluoto (2019), on the other hand, covers a wider time range and offers many excellent insights into the conceptual issues underlying the debate. Losee (2004) is also worth consulting to see how earlier accounts of what scientific progress consists in do not all fit into the classification system suggested by Bird (2007).

The exact scope of 'cognitive' is contested, however, which manifests itself in how different authors define the notion. Laudan (1977: 7) employs a broad definition, in taking cognitive progress to be 'nothing more nor less than *progress with respect to the intellectual aspirations of science*'. Niiniluoto (2019) instead introduces it as involving 'increase or advancement of scientific knowledge' or 'success in knowledge-seeking or truth-seeking'. And Dellsén (2018a) writes that 'cognitive scientific progress ... has to do with improvement in our theories, hypotheses, or other representations of the world'. In effect, Niiniluoto and Dellsén build some of their own views on what constitutes cognitive progress into their initial characterisations of the cognitive domain. To remain more neutral on the matter, let us instead begin by assuming only Laudan's broader, less presumptuous, definition.

The second limitation of this survey is its historical scope. Cognitive scientific progress has been discussed for as long as anything resembling modern science has existed; even before the twentieth century, it was discussed by Bacon, Whewell, Mill, Mach, Duhem, and Poincaré. It has also been addressed by many scientists, as illustrated in Rowbottom (2019: ch. 4). In the past century, moreover, interest in the topic blossomed after the publication of three classics: Hanson's *Patterns of Discovery*, Popper's *Logic of Scientific Discovery*, and Kuhn's *Structure of Scientific Revolutions*. All leading philosophers of science since – such as Cartwright, Kitcher, Laudan, and van Fraassen – have had something to say about progress, even if they didn't label their work as being on the topic. But since a full survey of this literature is beyond the scope of this Element, I have elected to focus on the current debate. Bird (2007) fomented this, and the bulk of the subsequent literature criticises his position and argues for alternative views with reference to his.

1.1.1 Key Preparatory Distinctions

Three preparatory distinctions will be useful. First, a complete answer to 'What constitutes scientific progress?' would concern two interconnected matters. On the one hand, it would specify which aspects of science are pertinent when evaluating cognitive 'goodness' in science. Should we limit our attention to theories? Or should we also consider existential claims? And should we be concerned ultimately only with beliefs concerning such things? What about other components of science, such as values, methods, exemplars, models, explanations, and instruments? In short, what are cognitive *goodness bearers*? On the other hand, it would specify on what basis we should compare those items to ascertain whether, or how much, progress has occurred. Imagine, for instance, that all goodness bearers are theories. Candidate cognitive *goodness makers* are potential properties of those goodness bearers, such as truth, approximate truth, empirical adequacy, simplicity, scope, and being known. Naturally, what goodness bearers are constrains what goodness makers can be, and vice versa. If truth is the sole goodness maker, for example, then methods are not goodness bearers (because they do not have truth values).

Second, note the distinction between *monistic* and *pluralistic* accounts of scientific progress. Monistic accounts hold that there is just one kind of goodness maker (although they often allow that there is more than one kind of goodness bearer). Pluralistic accounts hold that there are various kinds of goodness maker (but might posit only one type of goodness bearer). For instance, a pluralist might hold that theories are the only goodness bearers, but that their 'cognitive goodness' involves several dimensions, such as accuracy, simplicity, and scope.[4] Some accounts are *more pluralistic* than others, moreover, in positing more kinds of goodness makers than others. It is imperative to understand, however, that a pluralistic account may be *hierarchical*, nevertheless, in the sense that it may rank some goodness makers as more significant (or 'core') than others. For example, a pluralist might hold that increases in truthlikeness always bring more progress than increases in simplicity, while accepting that progress can occur either way.

Third and finally, heed the distinction between changes that constitute progress and changes that promote progress. The difference is sometimes easy to see. For example, drastically increasing universities' research funding would probably result in cognitive progress but would not itself constitute a cognitive 'good'. Part of what makes this obvious, though, is that increasing research funding is not a change of a cognitive or intellectual kind. When we instead consider changes internal to science which feature centrally in what scientists do – the development of instruments such as the scanning electron microscope (SEM) and lab techniques such as gene splicing, for example – it is harder to determine what we should say. A complicating factor is the possibility, especially from a pluralistic perspective, that some kinds of change might simultaneously promote progress and constitute progress. Consider again the development of the SEM. One might argue that this constituted cognitive progress by enabling us to do new things and promoted progress by leading to the discovery of new truths (about things investigated using the SEM).

Having presented these distinctions, I will now turn my attention to the dominant extant accounts of scientific goodness makers and goodness bearers, with a special focus on the most prominent defenders of each. In doing so, I will follow the thread of the debate initiated by Bird (2007) in a broadly

---

[4] Each of these items appears on the list of theoretical virtues proposed by Kuhn (1977: ch. 13).

chronological fashion. In the interests of economy, I will rarely hereafter use 'cognitive' to delimit the scope of the claims about progress.

I will cover the four dominant monistic (or near-monistic) accounts: epistemic (knowledge-based), semantic (truth-based), functionalist (problem-based), and noetic (understanding-based). Each has different variants. As we will see, pluralistic accounts tend to combine elements of each.

1.2 Bird's Epistemic Account: Progress As Increasing Knowledge

Bird (2007) reinvigorated the debate on scientific progress by championing a simple monistic view thereof: scientific progress occurs if and only if scientific knowledge increases. As Bird concedes, the idea is not novel. For instance, as Charkravartty (2017) notes, realists typically 'regard theories as furnishing *knowledge* of both observables and unobservables [and] arguably the most important strains of antirealism have been varieties of empiricism which, given their emphasis on experience as a source and subject matter of *knowledge*, are naturally set against the idea of *knowledge* of unobservables' [my emphasis]. That is to say, realists and anti-realists disagree about the *scope* of the knowledge science provides. But both sides typically agree that science provides some knowledge. It is therefore natural, prima facie, for both sides to accept that science makes cognitive progress when scientific knowledge increases.

However, Bird's epistemic account is bold because of its monistic character; no one else has defended such an epistemic account in detail.[5] This is partly why it is so interesting. It is also no accident that Bird's account came in the wake of Williamson's (2000) knowledge-first approach to epistemology, with which it dovetails. According to this approach, which is defended by Kelp (2014), finding knowledge is the constitutive aim of inquiry. And to this, to reach a position approximating Bird's epistemic account, one need only add that: (a) science is a form of inquiry; and (b) to achieve (or approach achievement of) the aim of an activity is the only way to make progress therein. As Bird (2007: 65) puts it: 'Given that science is an epistemic activity it seems almost tautologous to suggest that its success and so progress should be measured by epistemic standards.' In due course, however, we will see that many philosophers of science have resisted this seemingly 'almost tautologous' claim.

Bird (2007) proceeds by comparing his epistemic account with the most prominent prior accounts of progress. I will now present these comparisons and summarise the subsequent exchanges concerning these in the literature.

---

[5] There are, however, precursors that Bird (2007) does not cite. Cohen (1980: 491), for example, argued: 'in science ... the objective is not truth but knowledge'. See also Barnes (1991).

I will begin by using Bird's (2007) sketches of alternative accounts of progress. Some respects in which these are inadequate will emerge subsequently.

1.2.1 Bird on the Epistemic View versus the Semantic View

Bird (2007: 65) initially attacks the semantic account of progress, which he characterises as 'progress is the accumulation of true scientific beliefs [or] a matter of increasing verisimilitude … or nearness to the truth'. His concern is that this semantic view fails to save our intuitions concerning episodes where luck is involved in new true beliefs arising or in the verisimilitude of scientific beliefs increasing.

Bird uses one highly abstract thought experiment, and another which appeals to a historical episode. The former involves a 'scientific community that has formed its beliefs using some very weak or even irrational method M, such as astrology' (Bird 2007: 66). A natural reaction to such a thought experiment, however, is that a community fails to be scientific if its 'true beliefs are believed solely because they are generated by M' (66).[6] The subsequent literature has therefore focused on his other thought experiment, which involves a real scientific community. Here it is:

> Réné Blondlot believed in the existence of what he called N-rays for what it is clear were entirely spurious and irrational reasons. Imagine for sake of argument that we were to discover that there are in fact hitherto unobserved entities answering to Blondlot's description of N-rays. So Blondlot's belief in N-rays would have been true but unjustified and not knowledge. The semantic approach would have to regard Blondlot's belief (which was widely shared in France) as constituting progress. That is clearly wrong. (Bird 2007: 67)

However, Rowbottom (2008) argues that matters are not so clear, because the history of science shows that Blondlot's (and the community's) belief in N-rays was based on several other false beliefs.[7] Hence, he argues that the episode might be construed as *regressive overall* on the semantic view, even if N-rays exist, provided the introduction of those numerous false beliefs was detrimental enough to outweigh any improvement derived from the introduction of some true beliefs about N-rays. To hold this, one need only reject the crude notion that progress *only consists in the accumulation of true beliefs*. One should grant also

---

[6] For instance, Niiniluoto (2017) writes: 'the primary application of the notion of scientific progress concerns successive theories which have been accepted by the scientific community. Some sort of tentative justification for such theories is presupposed even by a radical fallibilist like Popper … Irrational beliefs and beliefs without any justification simply do not belong to the scope of *scientific* progress.' Rowbottom (2015a) and Dellsén (2021), on the other hand, argue that such 'tentative justification' falls short of the kind of justification required for knowledge.

[7] Rowbottom (2008) cites Lagemann (1977) and Nye (1980). The history also casts some doubt on whether the reasons were 'entirely spurious and irrational'.

that *avoiding false beliefs* is significant. This idea is old. One finds it, for example, in James (1896: §vii):

> There are two ways of looking at our duty in the matter of opinion, – ways entirely different . . . *We must know the truth*; and *we must avoid error* . . . they are two separable laws.

James (1896) notes also that eliminating a false belief need not involve substituting it with another belief. His discussion is richly suggestive of the consequence that, ceteris paribus, removing a false belief is progressive and introducing a false belief is regressive when it comes to doing 'our [epistemic] duty'. No modern defender of the semantic view disagrees with this sentiment (although not all think in terms of duties). And as we will see, Cevolani and Tambolo (2013) show that this consequence is 'built in' to the verisimilitude-based variant of the semantic view.[8]

Rowbottom (2008) continues by presenting several variants of the N-rays-based thought experiment, aimed at showing the significance of the value problem in epistemology. Broadly, the value problem concerns a question which Plato tackles in *Meno*: 'Why is knowledge more valuable than belief?' Rowbottom (2008: 278) argues that appeal to justification does not suffice to answer satisfactorily; he asks, rhetorically: 'Wouldn't we be better off having an unjustified true belief that "N-rays exist" and being neutral concerning whether there is any evidence to that effect rather than having a justified true belief that "N-rays exist" based on justified false beliefs in a great deal of evidence to that effect?' Rowbottom (2008) continues by suggesting that justification is only *instrumental* in achieving progress and hinting that knowledge may be in a similar boat.

Bird (2008) responds to Rowbottom's (2008) point about false beliefs as follows. First, he makes an empirical claim about what drives our response to his thought experiment: 'It takes a certain amount of reflection to see that Blondlot must have had some false beliefs, even if his theory is true . . . I do not think that this reflection is in fact present when we carry out the thought-experiment' (Bird 2008: 208). However, Rowbottom (2008) did not deny this claim; rather, he suggested that the thought experiment appears to tell against the semantic view only if one mistakenly thinks that the semantic view is committed to the claim that the N-ray episode was progressive overall.[9] Second, Bird (2008: 280) claims that the semantic view of progress doesn't entail anything about false beliefs, and therefore that Rowbottom is proposing a revision to it:

---

[8] Indeed, Bird (2007: 85) himself notes that 'the truth view of the aim of belief is typically modified, so that the aim of belief is characterised as the complex aim of achieving truth subject to the proviso that falsity is always avoided'.

[9] Plausibly, we should also not be interested in snap judgements. We should reflect on our responses to thought experiments, formulate arguments as a result, and reach judgements on that basis. See, for instance, Deutsch (2015).

to accommodate Rowbottom's response ... the defender of the semantic view
will have to make changes to (S) [the semantic view]. This is because (S) says
nothing about false belief; it does not yet say that an episode is not progressive if
it involves a considerable quantity of false beliefs. All it says is that there is
progress when there is an increase in true beliefs – which there is in my
hypothetical example.

Bird continues by suggesting that such revisions would be problematic, and that
the epistemic view requires no such machinations: 'one has to complicate
matters by saying that belief aims at truth and at avoiding falsehood ... The
epistemic view of the aim of belief and of progress can avoid all this' (280).

As Cevolani and Tambolo (2013) point out, however, it follows from Bird's
(2007: 65) own definition of the semantic view, on which progress may be 'a matter
of *increasing verisimilitude ... or nearness to the truth*' [my emphasis], that the
elimination of false beliefs may result in progress (and that the introduction of false
beliefs may result in regress). Verisimilitude 'represents the idea of approaching
comprehensive truth' (Popper 1963: 237). To illustrate, imagine a future in which we
have the comprehensive scientific truth. We have all the true claims in science's
domain of inquiry: truths about the fundamental constituents of the world, about the
laws governing the behaviour of those things, and so forth. Now consider how the
degree of verisimilitude of our science could decrease. We might forget some truths.
Or we might replace some true claims with false ones. And the latter kind of
regressive change might sometimes be worse than the former kind. Avoiding
falsehood is important. Thus, no revision is required to the semantic view for
Rowbottom's (2008) argument to go through. To return to Cevolani and Tambolo
(2013: 925):

> verisimilitude is a 'mixture' of two ingredients, truth and content. If truth
> were the only ingredient, then all truths, including the tautology, would be
> equally (and maximally) verisimilar; and, vice versa, if only content were
> relevant, then a plain contradiction would be closer to the truth than any other
> theory. Thus, devising highly verisimilar theories is a 'game of excluding
> falsity and preserving truth'. [quoting Niiniluoto (1999: 73)]

So the semantic view is pluralistic to a small extent; it admits two dimensions of
progress. Hence, Cevolani and Tambolo (2013: 930) opine that 'Bird's
attack ... is apparently based on a misunderstanding ... First of all, verisimilitude should not be conflated with approximate truth; and secondly, the accumulation of (approximate) truths does not guarantee increasing verisimilitude.'[10]

---

[10] On the second issue, Cevolani and Tambolo (2013: 931) explain that 'accumulation of (approximate) truths is neither a necessary nor a sufficient condition for increasing verisimilitude'. It is
not necessary because abandoning a false claim could serve that end. It is not sufficient because
adding a true (or approximately true) statement to a false hypothesis can result in a less

Cevolani and Tambolo (2013) add that Bird's N-ray thought experiment involves a scenario where *estimated* progress and real progress come apart, and that such cases are accepted as commonplace by advocates of the verisimilitude-based view of progress. Niiniluoto (2014) agrees with this verdict, labelling Bird's account of the semantic view 'incomplete and misleading', and adds that estimates of progress must be justified. It is therefore reasonable to conclude that Bird's (2007) thought experiment only refutes a simplistic variant of the semantic approach – dubious since James (1896) – which takes progress to occur only via accumulation of true or approximately true statements.

This brings us to Bird's (2008) response to Rowbottom's (2008) suggestion that justification is only instrumental for progress. Bird agrees that adding justification has a negative effect on progress in some contexts, but explains that this need not go for knowledge, provided one adopts an appropriate account of knowledge, such as Williamson's (2000). Bird (2008: 280) writes that the epistemic view 'says knowledge constitutes progress, and nothing short of knowledge. It does not imply that justification constitutes progress (or some weaker progress-like good); even less does (E) imply that justification (or even knowledge) will cause future progress'. Bird adds that Williamson's view of knowledge nonetheless entails that justification is not merely instrumental in value, because justification is necessary for knowledge. In short, knowledge entails justification (and truth) on Williamson's (2000) view.

Rowbottom should have anticipated this reply, because Bird (2007: 72) states that truth and justification are not, on his view, 'jointly sufficient for a new scientific belief adding to progress'. Rowbottom (2010) acknowledges this, although, as we will see, he uses this relationship to underpin a different attack on Bird's epistemic view. However, Rowbottom (2010) also points out that an epistemic view need not suppose a Williamsonian view of knowledge. Like the semantic view, the epistemic view has different variants. Justification's instrumental value tells against some such accounts.

In any event, Bird (2007) does not rely entirely on thought experiments. He argues separately that the semantic account has no advantages over the epistemic account. Most notably, he writes:

> The notion of verisimilitude lacks a worthwhile characterization in place of a definition. It is in less general use than the concept of knowledge. It is not

---

verisimilar hypothesis. Their example, which is artificial but nonetheless illustrative, is as follows. Imagine we have the false theory, *T*, that Mont Blanc's height is either 1,000 or 4,809 metres. If we now discover the true claim that 'Mont Blanc's height is either 1,000 or 4,810 metres' and we add this to *T*, we conclude that 'Mont Blanc's height is 1,000 metres'. We have been led away from *the truth* by discovering *a true claim*. The underlying issue here is that elimination of error, discussed previously, is important.

obviously explanatorily significant. And, most importantly, it is difficult to see how its use can be helpfully extended beyond the simple cases we do apply it to. (Bird 2007: 75)

No one has responded to the claim that verisimilitude is in less widespread use than knowledge and is of dubious explanatory significance. But a possible response is as follows: (1) even if a knowledge-based account of progress would be easier for non-specialists to grasp and 'get behind', this does not indicate it is correct; and (2) an adequate epistemic account will be more complex than first appearances suggest, because some folk conceptions of knowledge are unfit for the task, as Bird concedes in ruling out justified-true-belief accounts. This is just a programmatic sketch of a response, however; this aspect of Bird's argument deserves more attention than I can give it.

This brings us to Bird's claim that it is difficult to extend the verisimilitude account beyond simple cases (without it amounting to a cumulative view). He takes the most significant problem to be as follows:

> Let it be that a science adds to the set of its generally accepted beliefs just one new belief that is . . . closer to the truth. There is now not even an intuitive sense in which the science as a whole is now closer to the truth than it was – unless that sense is identical to the thought that this science includes *more* (approximate) truth. (Bird 2007: 75).

However, as we have already seen, advocates of the verisimilitude view of progress are not committed to the view that progress always occurs in such circumstances. Adding new approximate truths – which most consider to be falsehoods – is only a potential way to increase verisimilitude.[11] So there is no problem here for Bird's target; any appearance to the contrary stems from conflating approximate truth with verisimilitude. To reiterate, Cevolani and Tambolo (2013: 933) warn that 'no simple principle of the form "add to T a true, approximately true, or verisimilar, belief" can guarantee truth approximation through belief change . . . [Although] if T' is obtained from a true theory T by adding a new truth to it, then T' will be more verisimilar than T'. See Footnote 10 and Niiniluoto (1999: 201–3) for more detailed discussion of this point.

1.2.2 Bird on the Epistemic View versus the Functionalist-Internalist View

Bird (2007: 67) also launches an attack on 'the functionalist-internalist' view of scientific progress, which he associates with Kuhn, Laudan, and, to a lesser extent, Lakatos (1978). He calls it functionalist because it takes progress to occur when science performs the function of 'solving problems

---

[11] For an alternative view, though, see Rowbottom (2022).

[or puzzles]'. He calls it internalist because it requires that those internal to science – the scientific community – can identify whether progress has been made.

Bird begins by claiming that much mundane scientific knowledge – such as data on 'stars and comets, or on new species and habitats' (Bird 2007: 68) – is acquired without problem- or puzzle-solving. But the argument here is too quick. For instance, Kuhn (1970: 25) acknowledged the significance of such activities in normal science:

> [One focus of investigation is] that class of facts that the paradigm has shown to be particularly revealing of the nature of things. By employing them in solving problems, the paradigm has made them worth determining both with more precision and in a larger variety of situations. At one time or another, these significant factual determinations have included: in astronomy – stellar position and magnitude, the periods of eclipsing binaries and of planets; in physics – the specific gravities and compressibilities of materials, wave lengths and spectral intensities, electrical conductivities and contact potentials; and in chemistry – composition and combining weights, boiling points and acidity of solutions, structural formulas and optical activities. Attempts to increase the accuracy and scope with which facts like these are known occupy a significant fraction of the literature of experimental and observational science.

This passage is part of Kuhn's (1970: 25) attempt to 'classify and illustrate the problems of which normal science principally consists'. Hence, Kuhn (1970) holds that such activities *are* (kinds of) puzzle- or problem-solving (and may thus be progressive). Kuhn's use of 'facts like these are known' even suggests that acquiring knowledge (in some sense) could be a *part* of making progress on a problem-solving view of progress; indeed, I will develop this point later in this Element. Kuhn's perspective is nuanced and not so easily refuted; I will return to it at various junctures, although I do not have the luxury of presenting it fully here.

Bird (2007: 68) continues by claiming that 'Kuhn and Laudan do not think of solving a puzzle as involving knowledge, when knowledge is understood in the classical way as requiring truth.' He adds:

> It is clear that both Kuhn and Laudan countenance contributions to scientific progress that do not involve any knowledge. In Laudan's case this is because he thinks that we never have scientific knowledge – he accepts the pessimistic induction as sound. (68)

However, the claim concerning Laudan is a non sequitur. To see this, consider Laudan's (1981: 35–6) key claim in the paper where he is said to endorse the pessimistic induction:

the realist needs a riposte to the *prima facie* plausible claim that there is no necessary connection between increasing the accuracy of our deep-structural characterizations of nature and improvements at the level of phenomeno-logical explanations, predictions and manipulations ... (Equally problematic ... is the inverse argument to the effect that increasing experi-mental accuracy betokens greater truthlikeness at the level of theoretical, i.e., deep-structural, commitments.)[12]

So Laudan (1981) does not deny that we have scientific knowledge. He just expresses a substantial doubt that we can have scientific knowledge of 'deep-structural' or 'theoretical' aspects of nature. And this is compatible with our coming to know more at the phenomenological level.

An important lesson to draw from this is that being an anti-realist (or a realist) does not entail taking any particular view on what progress is (rather than on what it is not).[13] The realism debate centrally concerns the *extent* of what we can know via science, or the *kinds of truths (or approximate truths)* we should expect science to *reliably* supply.[14] Indeed, some anti-realists think that pro-gress is made by finding new truths. Van Fraassen (1980), for example, thinks success in science comes via achieving (or getting closer to achieving) empir-ical adequacy; he defines empirical adequacy in terms of truth in the observable domain.[15] Wray (2018) follows suit.

Similarly, finding scientific truths need not always involve finding truths about nature. As Rowbottom (2010: 247) explains: '[on] the semantic view, the kind of progress made may be the proper classification of (along with belief in) true propositions such as "T_1 has less scope than T_3 but has greater accur-acy", "T_1 is more simple than, but is otherwise equally as virtuous as, T_2", and so forth.' This indicates that the different accounts of progress are not always easy to disentangle. As we will see, some may potentially be reduced to (or subsumed by) others.

---

[12] The 'pessimistic induction' claims nothing about future theories. It is a challenge to the 'no miracles argument' that the success of a theory indicates its probable approximate truth, which is based on putative historical examples of empirically successful theories that were not approxi-mately true. As shown by Saatsi (2005: 1090–1):

it is an argument to the *timeless* conclusion that 'successfulness of a theory is not a reliable test for its truth'. As a matter of fact, in this conclusion no reference is made even to the probable falsity of any one theory of the current successful science; this conclusion would indeed hold even if the current theories were all likely to be true!

[13] Being an anti-realist correlates with *rejecting* some views, such as that progress occurs via learning new things about unobservable entities.

[14] For detailed treatments of what scientific realism is, see Chakravartty (2017) and Rowbottom (2018a; 2019: appendix).

[15] Indeed, Bird (2007) notes that van Fraassen (1980) could adopt a limited epistemic view of progress.

In fact, Laudan (1984: 51) holds that scientific progress cannot consist in moving towards a utopian goal, or a goal 'whose realization we could not recognize even if we had achieved it'. Laudan considers such goals irrational and requires that scientific goals be rational; *he is an internalist for this reason.* Moreover, Laudan claims *that finding true theories* is a utopian goal. But he does not consider *establishing some truths* to be a utopian goal. Laudan (1977: 127) states, in a passage Bird (2007: 69) quotes, that 'we can determine whether a given theory does or does not solve a particular problem'. It follows that we can determine truths of the form 'Theory T solves problem P' and 'Theory T does not solve problem P'. They are truths about scientific theories and problems. This prompts the question 'Would Laudan have accepted that we can sometimes recognise that we *know* such truths?' It is not easy to discern. Much would depend on the specific theory of knowledge in play. In any event, a functionalist-internalist *could* develop a view along these lines.

What explains the confusion surrounding Laudan's statements on truth? As McMullin (1979: 623) notes: '[in Laudan 1977] it is assumed that . . . progress resides mainly in appropriate choice between competing theories'. So Laudan (1977) assumes that goodness bearers are (almost always) theories, and his comments about truth not being a goodness maker should be understood only in this narrow sense. The lesson is that assumptions about what goodness bearers are, which are often only implicit in discussions of scientific progress, can lead to misunderstandings.

This brings us to Bird's (2007: 68–70) central argument against the functional-internalist view, which proceeds by way of a memorable thought experiment:

> Nicole d'Oresme and his contemporaries believed that hot goat's blood would split diamonds . . . If Oresme can solve his problem by coming up with a theory from which the splitting of diamonds by hot goat's blood is deducible, then Oresme has thereby contributed to scientific progress [according to the functionalist-internalist view] . . . While such a solution might reasonably have seemed to Oresme and his contemporaries to be a contribution to progress it is surely mistaken to think that this *is* therefore a contribution to progress . . . Furthermore, imagine that some second scholar comes along and proves at time *t* by impeccable means that Oresme's solution cannot work. Whereas we had a solution before, we now have no solution . . . that would mark a *regress*. But the correct thing to say is that the later scholar did indeed contribute to progress in a small way, by giving us knowledge that something previously thought to be true is in fact false.

No one has objected to Bird's conclusion from this thought experiment. Many have agreed with it. Dellsén (2016) endorses the same response to it and Shan

(2019) holds that it refutes Laudan's and Kuhn's perspectives on progress. Cevolani and Tambolo (2013) also express broad agreement with it.

On balance, I concur with Bird's verdict on Laudan's view. Specifically, Laudan's assumption that only theories usually bear goodness, coupled with his claim that progress occurs only when those theories solve perceived problems, precludes the possibility that progress occurs by identifying problem-solving errors (except in exceptional circumstances).

However, Kuhn's views are more subtle and deserve more careful consideration. He allows that problem- (or puzzle-)solving involves a whole host of activities, and that goodness can be borne by entities other than theories. Kuhn's (1970) disciplinary matrices (or constellations of group commitments) have many different components, including methods, values, and exemplars (of successful puzzle-solving). Moreover, Kuhn (1977) adds that theories may have a variety of virtues, such as accuracy, simplicity, scope, and consistency.

Prima facie, nonetheless, I grant that it is difficult to square Kuhn's views with the idea that progress can be made by identifying errors of the kind Bird points to. After all, Kuhn (1970: 80) claims that in *normal science* – or science under a disciplinary matrix – 'failure to achieve a solution discredits only the scientist and not the theory'. However, because problem solving involves so many different activities – 'determination of significant fact, matching of facts with theory, and articulation of theory' (Kuhn 1970: 34) – scientists may solve (or contribute to solving) one kind of problem by showing that a problem of another kind has not been solved. Furthermore, Kuhn (1977: 235) emphasises how important it is to identify insoluble problems, and how scientists strive to achieve this to bring about revolutionary change (albeit by doing normal science):

> the practitioner of a mature science ... tries ... to elucidate topographical detail on a map whose main outlines are available in advance, and he hopes – if he is wise enough to recognize the nature of his field – that he will someday undertake a problem in which the anticipated does *not* occur, a problem that goes wrong in ways suggestive of a fundamental weakness in the paradigm itself. In the mature sciences the prelude to much discovery and to all novel theory is not ignorance, but the recognition that something has gone wrong with existing knowledge and beliefs.

It follows that knowing what cannot be achieved via a paradigm (qua disciplinary matrix) is important in promoting progress, at the bare minimum. That's because spotting weaknesses – serious anomalies, insoluble problems – foments revolutionary science. Indeed, this is one of the main points of doing normal science for extended periods, on Kuhn's view. In Kuhn's (1970b: 247) own words, because 'exploration will ultimately isolate severe trouble spots, they [i.e., normal scientists] can be confident that the pursuit of normal science will inform them when and where they can most usefully become Popperian critics'. And this recognition takes the sting out of Bird's thought experiment against Kuhn. (Kuhn's view may also not be as 'internalist' as Bird suggests; I will return to this when I discuss Shan's (2019) functionalist view.)

This said, when a community starts questioning its theories – when normal science breaks down as a crisis of confidence occurs and extraordinary science begins – it is unclear in what respect this involves spotting previous mistakes, from Kuhn's perspective, because what counts as a legitimate puzzle/problem changes as a new disciplinary matrix emerges. As Bird (2007) points out, there therefore needs to be some way to weigh gains and losses, and it is unclear what this could be. Bird's (2007) epistemic view faces no such problem. But those who believe Kuhn (1970) is on the right track will see the problem as unavoidable because world views can radically change via scientific revolutions.

1.2.3 Bird on Knowledge of False Theories, Promoting Progress via Knowledge, and Understanding

Bird (2007) makes three further claims, which will become relevant in what follows. First, he argues that the adoption of new false theories may be progressive because significant things can be known *about them*. He writes:

> Let A(. . .) be a propositional operator whose meaning is given thus: A(p) iff approximately p. Assuming . . . that all the p_i are approximately true, the sequence of propositions $<A(p_1), \ldots, A(p_k)>$ will be a sequence of propositions each of which is fully true and adds to the truth provided by its predecessors. . . . The improving precision of our approximations can be an object of knowledge. (Bird 2007: 77)

Bird takes this to explain how much progress has been made in past science. In short, the important idea is that we typically do not come to know new theories directly. We could not, because most are false. Instead, we come to know items like 'T is approximately true' and perhaps (as suggested by 'improving approximations can be an object of knowledge') 'T_1 is closer to the truth than T_2'.

Second, he states that any cognitive changes which promote science's goal are progressive. (We will look closer at the talk of 'goals' or 'aims' in Section 2.) He suggests: 'one should regard science as progressing when a [cognitive] development promotes the growth of knowledge' (Bird 2007: 84). He continues, however, by stating that 'the relevant developments that promote knowledge will themselves be knowledge' (84). He claims that even changes in scientific methods will 'not typically be distinguishable from progress in scientific knowledge' (Bird 2007: 84).

Third, Bird (2007: 84) accepts the importance of understanding in science but claims that 'all understanding is also knowledge. To understand why something occurred is to *know* what causes, processes, or laws brought it about.' We will see how Dellsén (2016; 2021) and Rowbottom (2015a) challenge this in what follows.

1.3 Arguments for the Semantic View versus the Epistemic View

Rowbottom (2010) attacks Bird's epistemic view on three fronts. First, he argues that progress may involve unjustified changes (or Gettier cases). But since knowledge entails justification (or the absence of Gettier cases), on Bird's view, this means that progress need not involve an increase in knowledge. Second, he contends that knowledge is not necessary for stability (as Bird 2007 claims, following Williamson 2000). Third, Rowbottom (2010) argues that introducing false beliefs (and hence not knowledge) may promote progress. Let's address each argument in turn.

First, Rowbottom (2010: 242) again uses a thought experiment:

> Consider two planets; the inhabitants of the first have no justified scientific beliefs (although they have many true ones), whereas the inhabitants of the second have many justified scientific beliefs and considerable scientific knowledge. But the civilisations on each of these planets appear to be equally as advanced. Each has developed similar technology, has similar societies and institutions, uses similar scientific theories, and so forth. Roughly, we'd say they were at the same stage of technological development as we were at the turn of the twentieth century.
>
>   Now add that the people on each of the planets began in similarly primitive circumstances and proceeded without any outside interference ... If it is right to say that such a scenario is metaphysically possible, and if it is correct to say that the people on each planet have made some scientific progress (which I emphasise need not be equal), then the epistemic view of progress is refuted. That is, on the standard assumption that knowledge entails justification.

Rowbottom (2010) then argues that this scenario is metaphysically possible and that scientific progress is necessary for the level of technological progress made on the planets. The only challenge to this has come from Niiniluoto (2014: 4), who states that 'irrational beliefs and beliefs without any justification simply do not belong to the scope of *scientific* progress' (and who opposes Bird's thought experiments too as a result).[16] However, Rowbottom (2010: 242) notes that 'if the discussion is to be restricted to Williamson's view of knowledge (which Bird supports) ... it is possible to have justified true beliefs without knowledge; and therefore the comparison might be drawn between planets inhabited by people with justified true scientific beliefs and people with scientific knowledge'. Such

---

[16] As we will later see, another rejoinder is that several episodes covered by historians of science would count as non-scientific according to such a criterion – and that regress is sometimes the result of a lack of justification – although this is not a matter we can pursue here.

a modified thought experiment is immune to Niiniluoto's (2014) criticism, as Rowbottom (2015a: 103) emphasises.

Second, Rowbottom (2010) considers that one might object to his thought experiment by arguing that stable consensus is necessary for progress. And knowledge might be the source of such stability. Bird (2007: 83) writes: 'Precisely because knowledge is harder to achieve, it is more stable than true belief. On the semantic approach progress is too easy – it can be accidental, in which case regress can occur just as easily.' However, Rowbottom (2010) casts doubt on this. He pays special attention to the claim of Williamson (2000: 79) that 'If your cognitive faculties are in good order, the probability of your believing p tomorrow is greater conditional on your knowing p today than on your merely believing p truly today'. Rowbottom (2010) accepts that Williamson (2000) presents cases where knowledge would be more stable than true belief. However, he argues that those cases fail to support the probability claim because Williamson 'also needs to show that there are not (or are fewer) equivalent cases in which true belief is more stable than knowledge' (Rowbottom 2010: 250). Rowbottom (2010) adds that even if the probability claim were true, it is unclear whether the difference would be significant. What if the probability of believing p tomorrow if you know p is only slightly higher than that of believing p tomorrow if you just truly believe p, for instance? He also claims that his thought experiment only relies on there being some way to grant stability to true beliefs that falls short of knowledge.

Third, Rowbottom (2010) challenges Bird's (2007: 84) claim that 'developments that promote knowledge will themselves be knowledge'. He argues that false theories can reliably be used to derive some true results; for instance, one can determine that the best 'angle at which to throw a projectile on earth, while on the flat, in order to maximise its horizontal displacement . . . is 45 degrees' (Rowbottom 2010: 252). Rowbottom suggests that this might count as knowledge, despite being derived from a false belief, on some views thereof (such as reliabilist ones). Unfortunately, however, Rowbottom (2010) neglects to consider Bird's (2007) appeal to knowledge of approximations.[17]

This brings us to the final set of recent arguments for the semantic view as against the epistemic view. These stem from Niiniluoto (2014) and target Bird's

---

[17] Rowbottom (2015a: 104) offers other arguments to the same effect, which are better. For instance:

> Imagine a leading scientist lies about performing an experiment to test [a] theory . . . but if he *had* performed the experiment, he would have got (approximately) the same results as those he fabricated. The lies, and the resultant beliefs in the scientist's testimony, promote community belief in the theory. And then experiments to test the theory are performed, much sooner than they otherwise would have been. The theory is genuinely corroborated as a result! (And you are welcome to imagine this results in *knowledge* that the theory is genuinely corroborated, and thereby of the theory.)

(2007) strategy for allowing that science has made progress via endorsing sequences of false theories. Recall that this involves appeal to knowledge of claims of the form A(T), or 'It is approximately true that T', at the bare minimum.

Niiniluoto (2014) has three distinct lines of attack. First, he asks how we could know that statements like A(T) are true and adds that 'the problem of reliable estimation of truthlikeness or approximate truth reappears here' (Niiniluoto 2014: 76). He says no more on the issue.[18] But his point is presumably that any problems the semantic view has with showing how progress is to be estimated also arise for the epistemic view. Bird could respond by insisting that there is a link between empirical success and approximate truth, in line with the so-called no miracles argument of scientific realists; empirical success indicates (probable) approximate truth. However, as we will see, this would not suffice to answer Niiniluoto's final concern.

Second, Niiniluoto (2014: 76) argues that community beliefs in statements like A(T) do not fit the history of science: 'Scientists may formulate and even accept false theories without trying to specify such margins of error which would make them true or even probable, so that we don't find A(NEWTON) in the *Principia*.' Bird might respond that scientists do often believe things like A(T). After all, T entails A(T), so a rational person who believes T and considers A(T) is likely to form a belief in A(T) too. However, this response doesn't work because those kinds of beliefs wouldn't count as knowledge. One cannot come to *know* A(T) simply by justifiably believing the false claim that T and deriving A(T).

Finally, Niiniluoto (2014) argues that Bird's epistemic account cannot satisfactorily distinguish between regressive and progressive sequences of theory change. He asks us to imagine a series of theories – T_1, T_2, ... T_n – which are increasingly farther from the truth. He points out that this is consistent with each being approximately true. Hence, we could gain cumulative knowledge according to Bird, and progress according to Bird, via theoretical changes that seem regressive. A few responses are open to Bird, however.

On the one hand, Bird might insist that knowledge of A(T_1) & A(T_2) is a genuine improvement over knowledge of A(T_1), even if we happen to mistakenly happen to think that T_2 is nearer to the truth than T_1. For instance, comparisons between T_1 and T_2 and later theories might bear ultimately some fruit; by considering similarities and differences between different approximately true theories, one might get an insight into *which elements* of each are accurate. However, a reasonable rejoinder is that scientists do not typically keep old

---

[18] Niiniluoto (2014) intends there to be four lines of attack, but one involves asking: 'What would be A(H) for a theory H with mistaken existence assumptions (like phlogiston theory) or implicit counterfactual idealizations (like the ideal gas law)?' This is a special case of the problem discussed here.

dominant theories in mind. They do not usually use them to guide the generation of future theories, as they should do if they think they are approximately true.[19]

On the other hand, Bird might argue that scientists can come to know claims like 'T_1 is closer to the truth than T_2', as suggested by his comment, noted in Section 1.2.3, that 'The improving precision of our approximations can be an object of knowledge' (Bird 2007: 77). However, this claim is underdeveloped as it stands. It would also be somewhat surprising if Bird embraced this strategy when he is so doubtful of our ability to reliably ascertain the relative verisimilitude of statements except in simple cases. Nevertheless, an advocate of the epistemic view *might* take this approach.

1.4 Folk Intuitions on the Semantic and Epistemic Views

The many different thought experiments concerning the semantic and epistemic views have led to an impasse, and dissensus on the nature of cognitive goodness makers persists. However, Mizrahi and Buckwalter (2014) attempt to break the deadlock by testing folk responses to several thought experiments. Their key findings are that: (1) most participants judge it to be possible for progress to occur when justification is not present; although (2) most participants take the presence of justification to contribute significantly to progress. They conclude that the semantic view lacks some explanatory power, but not that the epistemic view is correct.[20]

Some have responded that we should not expect folk intuitions to be valuable data about progress, even assuming that philosophers' intuitions are useful; folk are typically not well versed in the history and philosophy of science. Moreover, it is arguable that philosophers of science rely on considered judgements regarding scenarios, not merely on immediate intuitions. For more on this issue, which we cannot delve into here, see Nado (2014).

1.5 An Argument for Extending the Epistemic View: A Role for Know-How?

Mizrahi (2013) proposes that we should look to what scientists say and think about progress. He uses two case studies involving Nobel Prize winners to do so. Specifically, he explores Landsteiner's discovery of blood groups and Pavlov's work on the physiology of digestion.

---

[19] As Rowbottom (2021) points out, one of the odd consequences of the view that we know lots of things like A(*T*) is that old theories become evidence about how new theories should look. But scientists do not use all their old theories – although they often use those of the last generation – in such a fashion.

[20] Rather surprisingly, Mizrahi and Buckwalter did not suggest a third route. Perhaps progress can be constituted by increases in verisimilitude *and* increases in justification – or even the achievement of knowledge – in addition?

He finds that scientists often describe progress in terms of knowledge and argues that the conception of knowledge they have is broad. Specifically, he argues that it not only covers theories and observations, but also covers:

> (PK) *Practical knowledge*: Practical knowledge usually comes in the form of both immediate and long-term practical applications.

> (MK) *Methodological knowledge*: Methodological knowledge usually comes in the form of methods and techniques of learning about nature. (Mizrahi 2013: 380)

Mizrahi (2013) claims that both PK and MK involve *know-how* – for example, how to study the anatomy of conscious animals – and not *knowledge that*. Thus, he questions the narrow view that the semantic and epistemic accounts take on what goodness bearers are (and not only on what goodness makers are). He quotes Baird and Faust (1990: 147) approvingly: 'Technicians, engineers, and experimenters . . . are able to make devices work with reliability and subtlety when they can say very little true, or approximately true, about how their devices work . . . Their knowledge consists in the ability to *do* things with nature, not *say* things about nature.'

The idea that *knowledge how* is distinct from *knowledge that* derives from Ryle (1949). For example, it is dubious that knowing how to ride a bicycle consists purely in having propositional attitudes, such as beliefs, because we cannot even come close to listing the relevant propositions. It is also implausible that anyone could learn how to ride a bicycle without mounting a bicycle (or something similar), but instead by learning that many propositions are true from a book. And although some have argued that *knowledge how* is a species of *knowledge that* – Stanley and Williamson (2001) maintain that knowing how involves entertaining propositions in a practical, rather than a demonstrative, mode of presentation – strong arguments remain against such a view. As Wallis (2008) illustrates with appeal to empirical psychology, we often explicitly deny facts about how we do things. Hetherington (2011) goes even further and argues that *knowledge that* can be reduced to *knowledge how*.

A standard criticism of Mizrahi's (2013) approach is that scientist's opinions are not good indicators of what scientific progress consists in.[21] For instance, Niiniluoto (2019) notes:

> scientists often have different opinions about the criteria of good science, and rival researchers and schools make different choices in their preference of theories and research programs. Therefore, it can be argued . . . that progress should not be *defined* by the actual developments of science: the definition of progress should give us a normative standard for appraising the choices that the scientific communities have made, could have made, are just now making, and will make in the

---

[21] Mizrahi has continued to study the views of scientists, however. See especially Mizrahi (2021).

future. The task of finding and defending such standards is a genuinely philosophical one which can be enlightened by history and sociology but which cannot be reduced to empirical studies of science.

This passage raises some profound questions about what the philosophers involved in the debate on goodness makers are doing. For example, are they *discovering* a normative standard or *imposing* one? We will turn to such questions in Sections 2 and 3.

Mizrahi (2013) takes the burden of proof to be on those who think changes in *know how* are not progressive. One line of counterargument he does not consider, however, is that increases in *know-how* only promote, but do not constitute, progress. In other words, developing new instruments like MRI scanners, and new techniques like gene splicing, might enable (or increase the probability of) the attainment of new cognitive goods – for example, new *knowledge that* – without *being* cognitively good. The prospect of easily settling the case appears dim, because there is considerable dissensus on this matter in epistemology, as there is in philosophy of science. However, the notion that *know-how* constitutes a distinctive kind of cognitive achievement is respectable; Carter and Pritchard (2015) defend it. Rowbottom (2019) also suggests that *know-how* is involved in prediction because this is not always a straightforward matter of plugging numbers into an equation; it often involves modelling and mathematical practices. Shan (2019) also thinks it is important, as we will see.

A reminder is appropriate, before we continue. As noted in Section 1.2, some authors make assumptions about the scope of 'cognitive progress' – and thus, implicitly delimit what goodness bearers can be – that lead them not to consider know-how. For example, Dellsén (2018a: 2) claims cognitive progress concerns only 'improvement in our theories, hypotheses, or other representations of the world, rather than other improvements of or within science'. This is unfortunate because it makes it look as if some other philosophers who purport to be discussing cognitive progress should be construed as intentionally discussing another dimension of progress. For instance, Chang (2004: 227–8) understands cognitive progress in a highly pluralistic fashion, according to which 'the enhancement of any feature that is generally recognised as an epistemic virtue constitutes progress'.

1.6 Arguments for an Understanding-Based (and Prediction-Based) View

Several authors – Bangu (2015), Rowbottom (2015a; 2019), and Dellsén (2016; 2021) – have also argued that increases in understanding may constitute progress, although there are differences in the way they construe 'understanding' and in the arguments they advance.

1.6.1 Understanding as Progressive from a Pluralistic Perspective

Rowbottom (2015a) takes aim at a narrow verisimilitude-based version of the semantic view, which is that progress occurs solely when theories increase in verisimilitude. Rowbottom's (2015a) case rests on yet another thought experiment. He asks us to imagine that scientists discover a maximally verisimilar theory for their field and recognise it as such. He points out that no more scientific progress would be possible in that area, on the view he is attacking. However, he argues that this is wrong on two counts:

> First, the true theory could be difficult, or even impossible, to use for predictive purposes. It could concern some initial conditions that are beyond our ability to determine the values of. Or it could be unusable in many situations in which predictions would be desirable, due to the need for arduous calculations. Second, even if it were of considerable predictive use, it might fail to measure up to our explanatory expectations. Imagine that it involved considering numerous variables, such that it was hard to appreciate how changes in one would tend to affect changes in another, in many applications, without the use of extensive computer simulations. It would not serve to grant insight. (Rowbottom 2015: 101)

In summary, the idea is as follows. Increasing predictive power or understanding is intuitively progressive. However, having a true theory doesn't mean being able to employ it to predict in all the ways one would like to. Nor does it entail being able to fully grasp the theory, or how it says various phenomena interrelate. Rowbottom (2019: §1.1) uses a case study involving pendulum models to illustrate the latter point. He points out that increasing the accuracy of such models involves introducing extra complexity, which makes them harder to comprehend and obscures some of the relationships they capture. He adds that the simple model of the pendulum makes it easy to see how (and when) pendulum motion is approximately simple harmonic; it also makes it easy to see how changing a pendulum's length, or the strength of its surrounding gravitational field, affects its swing frequency. But Rowbottom then shows how difficult it is to grasp a complex model that includes frictional forces. He concludes that if we started with a highly accurate model, we might want to simplify it by idealising, and making it less representationally accurate, to increase our understanding. This also fits with what others have argued; for instance, Rancourt (2017) holds that increases in false belief can further understanding.[22] Thus, Rowbottom's (2019: ch. 5) view of understanding is that it does not require accurate representations (and hence anything like

---

[22] See also Potochnik (2017).

approximate truth or knowledge). Rather, for Rowbottom (2019: 3), scientific understanding 'involves representations that are cognitively appropriate to serving scientists' empirical ends, in so far, for example, as they are highly memorable and easy to use'.

Unfortunately, how we should construe 'understanding' is hotly contested. The literature on the topic has exploded over the past decade, and I cannot cover even a tiny fraction of this.[23] For present purposes, I can only point out how understanding is construed by those who have argued, in the debate covered here, that it contributes to progress.

1.6.2 Understanding As Progressive from a Monistic Perspective

Dellsén (2016: 72) proposes a monistic account according to which: 'an episode in science is progressive precisely when scientists grasp how to correctly explain or predict more aspects of the world at the end of the episode than at the beginning'. Dellsén notes that it is unusual to take understanding to cover prediction but suggests that 'a scientist who makes correct predictions about something . . . without also realizing how to explain it does seem to have at least *some* understanding of the phenomenon in question' (Dellsén 2016: 75). On Dellsén's view, therefore, understanding is a matter of degree and partial understanding is present in such cases; he holds that increases in the *extent* of understanding may constitute progress. Specifically, he suggests: '(U) An agent has partial scientific understanding of a given target just in case she grasps how to correctly explain and/or predict some aspects of the target in the right sort of circumstances' (Dellsén 2016: 75).

Dellsén (2016) concedes that many would object to his theory of understanding. But such individuals might still agree with the core of his account of progress. Thus, he is content for (U) to be understood as a stipulative definition. This means taking his account to be pluralistic: to say that progress occurs if and only if scientists grasp how to do one of two different things – predict or explain – better or more.

Dellsén (2016) uses Bird's epistemic (2007) view as his foil; he argues that his own noetic view is superior. He therefore begins by explaining how partial understanding – as defined in (U) – is possible in the absence of justification or belief (and hence also knowledge) on his view. For instance, he holds that scientists sometimes accept theories as true, rather than believing in them, yet achieve understanding by deploying them appropriately.

---

[23] This includes several monographs, most notably de Regt (2017) and Elgin (2017).

Dellsén then argues for two theses, with appeal to historical examples and thought experiments: (1) increases in scientific understanding are possible without increases in scientific knowledge (concerning theories or phenomena); and (2) some increases in scientific knowledge are not progressive because they do not also involve increases in scientific understanding. Let's consider each argument in turn.

1.6.2.1 Progressive Understanding without Knowledge

Dellsén (2016: 76) asks us to consider a paper in which Einstein uses the kinetic theory of heat to explain Brownian motion. This is an interesting case, Dellsén argues, because Einstein knew neither the explanandum nor the explanans at the time of writing; he lacked the justification necessary to know that the kinetic theory of heat was true or that Brownian motion occurs. Nonetheless, Dellsén (2016: 76) argues: 'it seems clear that Einstein's explanation constituted significant (cognitive) progress in science. The noetic account explains why. On this account, Einstein made scientific progress because he enabled us to grasp how to correctly explain Brownian motion'.

Dellsén then considers two possible objections to this view. On the one hand, Bird (2007) could deny that progress occurred when Einstein's paper was published. But Dellsén rules out this option swiftly because Einstein's explanation of Brownian motion is widely accepted to have been highly progressive. On the other hand, Bird could argue that Einstein's paper provided the knowledge that kinetic theory was a *potentially correct explanation* of Brownian motion. Cleverly, however, Dellsén (2016: 77) points out that Bird seems to reject this possibility, in his thought experiment involving the generation of a theory for why hot goat's blood would split diamonds. Bird (2007: 69) writes, recall, that 'While such a solution might reasonably have seemed to Oresme and his contemporaries to be a contribution to progress, it is surely mistaken to think that this *is* therefore a contribution to progress.' But if there was no knowledge of a potentially correct explanation generated in the hypothetical case involving Oresme, then no such knowledge was generated in the case involving Einstein.

Overall, this is a strong line of argument. But perhaps Dellsén dismisses the first kind of objection too quickly. Consider the following possible rejoinder from Bird: Einstein's contribution did not constitute progress *in isolation*, although it was a core element of a progressive episode (and is celebrated as such). When Brownian motion was subsequently found to occur, the paper led to knowledge of how Brownian motion comes about and maybe also of the kinetic theory of heat. Furthermore, the paper may have led to progress even if Brownian motion had been found not to occur, because it might have resulted in knowledge that the kinetic theory of heat was false. Finally, Einstein's contribution also promoted progress because it encouraged exploration of whether Brownian motion occurs.

This response has potential, although it remains natural to think that Einstein's contribution *did* give us knowledge of how to confirm or disconfirm an active theory (by finding whether Brownian motion occurs). Bird could suggest that this is not the kind of knowledge that constitutes, rather than possibly promotes, progress; but then his account would need to be revised because the kind of knowledge involved appears scientific. Certainly, Einstein's paper was scientific.

1.6.2.2 Knowledge without Progressive Understanding

Dellsén also argues that there are several ways in which scientific knowledge can increase without any progress occurring. He covers three kinds of cases. The first involves the collection of data known to be randomly produced. The second concerns the discovery of statistical correlations between causally unrelated states of affairs, such as 'increases in childbirth rates outside of Berlin city hospitals and increases in stork populations around the city' (Dellsén 2016: 78). The third involves the collection of highly limited observation statements, such as 'a list where each entry has the following form: "Seagull observed at [time] on [date]"' (78) Dellsén points out that these cases are not progressive on the noetic view, because such findings have no predictive or explanatory value.

Dellsén begins by arguing that increases in knowledge occur in these cases. This is reasonably uncontroversial; for instance, the correlation he mentions is genuine. He then argues that a proponent of the epistemic view of progress should not accept that gathering such information is progressive, because 'a scientific practice organized around accumulating trivial knowledge of this kind would seem to be a paradigm example of degenerate science' (Dellsén 2016: 79). He concludes that the only workable alternative would be to concede that only *some* increases in scientific knowledge are progressive.

There is a more promising riposte, however, which Dellsén (2016: f. 30) hastily dismisses in a footnote. The knowledge gathered in such cases might not be scientific, even if scientists obtain it in the course of their work. Not everything a scientist comes to know while working constitutes progress on the epistemic account; the colour of a lab door, the number of radiators in an office, and so forth, are not items of scientific knowledge. And maybe the cases Dellsén presents involve knowledge acquisition of an equally trivial, non-scientific kind. For example, when a scientist runs a statistical analysis of a massive amount of data to spot correlations therein, she expects some findings that are not scientifically relevant according to her current theories. And she will typically discard those outputs where the correlation does not, on current theory, demand any kind of causal explanation. For illustration, consider Popper's (1963: 43) witty tale of 'the man who dedicated his life to natural science, wrote down everything he could observe, and bequeathed his priceless collection of observations to the Royal Society to be used as evidence. This story should show us that though beetles may profitably be collected, observations may not'. Popper's point was that the theoretical background determines what counts as a *scientific* observation statement, rather than a non-scientific one.

1.6.2.3 Public Understanding

Dellsén (2021) extends the noetic account in various respects, not all of which we can cover. One of his most interesting arguments is that authors on progress have been too focused on scientists' attitudes. He writes:

> How could the extensive funding of 'pure' scientific research, with no clear practical benefits for non-scientists, be justified if scientific progress merely consisted in some scientists improving *their* cognitive attitudes? In light of this problem, I suggest that we move to a conception of scientific progress according to which ... progress is determined by the publicly available information, such as that contained in peer-reviewed journal articles, on the basis of which any relevant member of society ... can form or sustain the relevant type of cognitive attitudes. (Dellsén 2021: 11257)

Naturally, one could consider a similar alteration to the competing accounts; one could take progress to depend on members of the public gaining access to more truthlike beliefs about nature, on the semantic view, for instance. But would this be right?

It is counter-intuitive that if the world's governments cooperated to prevent public access to all scientific papers, then all scientific progress would cease. It seems, rather, that science could continue to progress, behind closed doors, irrespectively. It appears too that the public might enjoy the benefits of such progress. For example, they might gain access to new medical treatments that would not have been possible without the scientists' efforts.

Dellsén does not consider this line of objection, but he might reply that he is *proposing* a standard for science in its current social context. A sensible compromise would be to allow that although science might progress, in the previous thought experiment, the governments' actions would constrain it.

1.7 Functionalist Responses to Bird's Arguments

Shan (2019) accepts Bird's critique of the functionalist-internalist view as represented by Laudan and Kuhn, but notes that Bird's taxonomy of views on progress is incomplete. For instance, why not adopt a non-internalist but nevertheless functionalist view of progress? Shan (2019: 744) proposes such as view, which he summarises as: 'Science progresses if more useful research problems and their corresponding solutions are proposed.'

He aims to distance his view on progress from the 'Kuhn-Laudan' approach in two main respects. First, Shan (2019: 744) claims that 'both Kuhn and Laudan implicitly assume that ... problems are either simply pre-defined or defined in a straightforward way ... [whereas] I argue that science consists of both problem-defining and problem-solving activities'. Shan says also that problem-defining involves three distinct activities: problem-proposing, problem-refining, and problem-specification. Second, Shan (2019: 745) states that 'Kuhn suggests that puzzle-solving is an activity of looking for a solution that is sufficiently similar to a relevant paradigmatic problem-solution ... In contrast, I do not think that the constituents of the solution to a research problem can be characterized in a monistic way.'

However, Kuhn would have strenuously objected to both claims. (I will focus on Kuhn because, as noted in Section 1.2.2, his position was especially sophisticated.) In response to the first, Kuhn (1970) argues that *most* problems (or puzzles) only emerge from disciplinary matrices, but also that disciplinary matrices only emerge from extraordinary science. That's to say, problem definition is a significant function of extraordinary science. Kuhn does not suggest that this is 'straightforward'. On the contrary, Kuhn (1970: 87) says the extraordinary scientist will: 'often seem a man searching at random, trying experiments just to see what will happen, looking for an effect whose nature he cannot quite guess'.

A potential retort is that for Kuhn, problems/puzzles are predefined in normal science. But Kuhn (1970: ch. 3) allows that problem-refining and problem-specification occur *in normal science*. He describes such activities as *articulation*. Articulation 'can resemble exploration' (Kuhn 1970: 29) and sometimes arises because 'a paradigm developed for one set of phenomena is ambiguous in its application to other closely related ones.' The following passage shows what Kuhn had in mind:

> Before he could construct his equipment and make measurements with it, Coulomb had to employ electrical theory to determine how his equipment should be built. The consequence of his measurements was a refinement in that theory. Or again, the men who designed the experiments that were to

distinguish between the various theories of heating by compression were generally the same men who had made up the versions being compared. They were working both with fact and with theory, and their work produced not simply new information but a more precise paradigm, obtained by the elimination of ambiguities that the original from which they worked had retained. In many sciences, most normal work is of this sort. (Kuhn 1970: 33–34)

If the paradigm (or disciplinary matrix) is the source of the problems, then when the paradigm is made 'more precise', the problems it 'defines' may thereby be refined (e.g., be made more specific). Kuhn also uses 'refinement' in writing of Coloumb's theoretical achievement; and as theories are parts of disciplinary matrices, refining these may also reasonably be expected to refine (a proper subset of) the available problems/puzzles.

This brings us to Shan's (2019) second claim, that 'puzzle-solving is an activity of looking for a solution which is sufficiently similar to a relevant paradigmatic problem-solution' on Kuhn's view. This also unfairly oversimplifies Kuhn's (1970) image of science. Partly, this is for reasons mentioned above; specifically, articulation need not involve looking for a solution of the kind suggested by some 'paradigmatic problem-solution'. To reiterate, Kuhn covers several different types of problem and problem-solving, both empirical and theoretical, in normal science. But he also holds that some problems occur *in extraordinary science*. Such problems need not have been 'pre-defined', let alone have a 'paradigmatic problem-solution', because they occur in the absence of any dominant paradigms. Here, again, is textual evidence:

> three classes of problems – determination of significant fact, matching of facts with theory, and articulation of theory – exhaust ... the literature of normal science, both empirical and theoretical. They do not, of course, quite exhaust the entire literature of science. There are also extraordinary problems, and it may well be their resolution that makes the scientific enterprise as a whole so particularly worthwhile. But extraordinary problems are not to be had for the asking. They emerge only on special occasions prepared by the advance of normal research. (Kuhn 1970: 34)

Note also that Kuhn (1970) only explicitly endorses internalism in the context of normal science. He strongly suggests that extraordinary science can involve appeals to criteria that are not internal to science in any strict sense, in the following passage:

> since no paradigm ever solves all the problems it defines and since no two paradigms leave all the same problems unsolved, paradigm debates always involve the question: Which problems is it more significant to have solved? Like the issue of competing standards, that question of values can be

answered only in terms of criteria that lie outside of normal science altogether, and it is that recourse to external criteria that most obviously makes paradigm debates revolutionary. (Kuhn 1970: 110)

Admittedly, 'criteria ... outside of normal science' might still be limited to criteria from scientists, but Kuhn does not specify this. In any event, my overarching point is that Kuhn's view of progress has not been considered carefully and charitably enough in the recent debate. It suffers from some problems. For instance, Kuhn's division between 'normal' and 'extraordinary' science is too sharp, and research may be possible under partial disciplinary matrices (as argued by Rowbottom 2018b). Perhaps, as Douglas (2014) argues, Kuhn also draws too sharp a line between pure and applied science. Nevertheless, his account has the resources to answer most existing criticisms of his views on progress.

Let us return to Shan's (2019) alternative. It employs the notion of 'usefulness'. But what is 'useful'? Shan (2019: 746) explains: 'A problem and its solution is useful if and only if the way of defining and solving research problems is repeatable and provides a reliable framework for further investigation to solve more unsolved problems and to generate more testable research problems across additional different areas (or disciplines).'

This passage is hard to follow. Partly this is because the 'if and only if' clause appears to refer to both problem and solution simultaneously (on the left-hand side), thereby suggesting that there are no useful problems that are not solved (or at least soluble). Moreover, to label a problem 'testable' is a category mistake. Theories or hypotheses are testable or untestable. Problems are soluble or insoluble. It is also unclear why problem solutions must be repeatable to be useful, even if repeatability is often a virtue. (Consider how solving a one-off problem in extraordinary science might bring about progress on Kuhn's view.) Shan (2019) gives the example of the Mendelian approach, although this appears to be a practice, bound up with various theoretical machinery. Overall, the ontology at play here is opaque; the account lacks the perspicuity and subtlety of Kuhn's. However, the basic idea should be clear.

Shan (2019) understands 'problems' to involve various practices. He follows (but does not cite) Mizrahi (2013) in thinking that increases in *know-how* are progressive: 'The way of defining and solving the problems is a clear case of know-how' (Shan 2019: 752). He also suggests that acquiring truths and propositional knowledge can be progressive: 'proposing more useful problems and their solutions could be interpreted as attaining more perspective-dependent true knowledge claims' (752). However, he only takes these to be two ways in which 'usefulness' can manifest; he also opines that the 'accumulation of scientific knowledge or the approximation of scientific truth is usually a result rather than an indicator of scientific progress' (Shan 2019: 754). His account is neutral on the realism debate.

Overall, Shan (2019) draws attention to several ideas relevant to articulating functionalist views: *know-how* is involved in problem-solving, problem-defining is significant, and internalism need not be presupposed. However, his view is not well-developed enough to be considered on a par with Kuhn's, which is seriously underestimated in the current debate. Functionalists would do better to accurately lay out Kuhn's account of progress, isolate the few extant objections that pertain to this, and employ the findings to generate new problem-solving views.

1.8 A Sprawling Interminable Debate

The literature on progress has been so extensive, even over the past fifteen years, that I have not been able to cover it all. I have had to make some difficult decisions about what to omit. I have not covered Mizrahi (2017) on a putative tacking problem with the semantic view or the response by Cevolani and Tambolo (2019). Nor have I covered Park's (2017) objection to the noetic view or Dellsén's (2018b) reply. I have not even covered my own pluralistic (but hierarchical) account of progress in Rowbottom (2019: ch. 1).

I have also not discussed interesting work with a narrower focus, such as Saatsi (2019) on how theoretical progress might be construed from a 'minimal realist' perspective. Finally, I have not been able to connect work on 'the aims of science' – such as Potochnik (2017) – with the discussions explicitly focused on scientific progress. But as the next section will show, there *is* a connection.

Similar new work on scientific progress appears apace. For example, Emmerson (2022) launches a new assault on Dellsén's noetic view, Bird (2022) defends his epistemic view of progress in a monograph, and Shan (2022) brings together a variety of new papers on the subject.

1.9 Conclusion: The Relevance of Meta-Normative Concerns

'What constitutes scientific progress?' continues to be answered in strikingly diverse ways by different authors. Moreover, novel proposals on the topic appear continually. But there is little consensus on the defects of the older accounts of goodness makers and even on the details of those accounts. We have more potential answers to the question than ever before, yet do not agree on a principled way to choose between them.

This situation suggests that the time is ripe to consider the foundations of the debate. Inter alia, we might ask the following. Were we right to presume that there is a timeless, context-invariant answer to what scientific progress consists in? And what about the world, if anything, makes one kind of thing a goodness maker, rather than another? What, that's to say, are the ultimate grounds for one kind of change being an improvement and another not?

Tackling such meta-normative questions may help in several respects. First, we should consider whether we have been on a wild goose chase. Have we been trying to answer a question like 'What's the best flavour of ice cream?' And if not, why not? Second, if we can establish what (ontologically) *would* make some things rather than others count as improvements, then this may help us to agree on how to answer our question. Are thought experiments useful? Or should we only look to what scientists are (and have been) striving to do? If we can agree on our ultimate target and methods to study it, then a coherent research programme might emerge in place of the disorderly exchanges still taking place.

There is also a nascent debate on what constitutes philosophical progress. Lessons learned about scientific progress could prevent wasted effort in this highly similar context, as argued by Dellsén, Lawler and Norton (2022).

These are my primary motives for considering foundational meta-normative issues concerning scientific progress in the remainder of this Element. But even if I am wrong about the inauspicious state of the current debate, these issues deserve attention because they are sorely neglected in the existing literature.

2 On Second-Order Cognitive Goodness Makers: The Aim(s) of Science

> The view that science has an aim or aims qualifies as a *bona fide* dogma in the philosophy of science. (Resnik 1993: 225)

> Never use a metaphor, simile, or other figure of speech which you are used to seeing in print. (Orwell 1946: 264)

Progress is a normative notion, as we have already seen. In Niiniluoto's (2019) words: 'to say that a step from stage A to stage B constitutes progress means that . . . B is *better* than A relative to some standards or criteria'. But whence do such standards or criteria originate? What *makes* something a pertinent standard or criterion?[24] To ask such questions is to inquire into the nature of *second-order* goodness makers; it is to seek *what makes a (first-order) goodness maker a goodness maker.*

---

[24] I cannot provide a detailed discussion of how standards relate to rules or norms. However, I presume only the following. First, standards are involved in evaluations. Second, prescriptions depend on evaluations (and thus involve the same standards). Consider an evaluation such as 'Science will make more progress if we do X than it will if we do Y.' This could be grounds for a prescription: 'Do X rather than Y if these are the only two options.' In such circumstances, the prescription results from the standards. (The link is not straightforward, because doing X could be supererogatory, but the idea should be clear. Evaluations have *deontic significance*, one might say.)

The standard answer is 'the aim(s) of science', as illustrated by the following quotations:

'Progress is always "progress relative to some set of aims".' (Laudan 1984: 66)

'A philosophical analysis of scientific progress is tantamount to a specification of the aims of science.' (Niiniluoto 1984: 76)

'Our conception of scientific progress is linked to what we take the aim of science to be. In general, something like the following principle holds: (A) if the aim of X is Y, then X makes progress when X achieves Y or promotes the achievement of Y.' (Bird 2007: 83–4)

'(A) X is the aim of science just in case science makes progress when X increases or accumulates.' (Dellsén 2016: 73)

It follows that identifying science's aim(s) is sufficient to identify what's scientifically progressive.

Mizrahi (2021) adopts an approach that seems, prima facie, to be of this kind. He presents a survey of what scientists say about their aims to 'contribute to the debate on the nature of scientific progress'. But understanding *what it is to be an aim of science* is crucial for evaluating the extent to which his study helps to determine what (first-order) goodness makers are. Imagine, for instance, that the aims of science are *not* aggregates of the aims of current scientists. Then scientists' immediate stated goals may be of minimal significance in assessing what constitutes scientific progress, because scientists typically work in narrow areas, perform specific functions, and rarely have training in philosophy. That's to say, their stated goals will tend to involve directed local tasks, such as seeking new species or determining the physical properties of new materials. Similarly, following van Fraassen (1994), we would not expect Clausewitz's view on the aim of war – that war is simply a vehicle for political ends – to derive support from soldiers' aims in concrete combat scenarios or soldiers' opinions about the point of war.

More strikingly, if science is aimless, as Rowbottom (2014) suggests, then scientific progress may not exist (in the objective or intersubjective fashion that most authors take it to exist).[25] And in that event, assertions of the form 'Science progresses when it achieves X' might be construed merely as expressing the utterer's positive attitude towards scientific change(s) resulting in X (in line with expressivist views in meta-ethics). Similarly, to allow that an aim of science is a shared aim of a quorum of scientists is to allow that what constitutes progress is contingent and can change over time. Yet all the most influential

---

[25] Rowbottom (2019: 188–9, f.6) uses 'the value of science' interchangeably with 'what's scientifically progressive' as a result of this concern.

contemporary views on the nature of scientific progress are monolithic: they offer contextually invariant accounts of what amounts to scientific progress.

In the remainder of Section 2, I survey the existing ways to construe 'an aim of science' and illustrate why they are problematic. I then propose that we consider hypothetically rational aims in doing science instead, to develop an account of second-order goodness makers. In Section 3, I build such an account.

2.1 Aims of Science: A Vague Dogma

Claims that science has an aim or aims feature in many of the most influential works in the philosophy of science. For example: 'science aims' plays a central role in van Fraassen's (1980: 12) definitions of scientific realism and constructive empiricism; 'aims of science' is prominent in Laudan's (1977; 1984) discussions of scientific change and the rationality of science; 'the aim of science' appears in the title of one of Popper's monographs; 'the aim [or object] of physical theory' features in another book by Duhem; and more recently, 'the aims of science' has titular significance in the work of Potochnik (2015; 2017). But what does it mean to say that science aims at something? What is it to be an aim of science?

Resnik (1993: 225) found no clear answer to this question:

> The view that science has an aim or aims qualifies as a *bona fide* dogma in the philosophy of science. Even though this seems like a reasonable (or at least popular) view, it is not at all clear how we should understand this view. After all, science *qua* science cannot literally have aims (ends, purposes, or goals); only beings with intentions (i.e., people) can have aims. So the phrase 'aims of science' must be taken as a metaphor.

The matter has rarely been discussed since. There was only an inconclusive exchange between Rosen (1994) and van Fraassen (1994) on the topic, which we will touch on later. And as Rowbottom (2014) argues, philosophers talk past one another by interpreting the phrase differently; some even equivocate when using it in their own work. Remarkably, no contemporary paper on progress mentioning 'aims of science' says, even roughly, what these are supposed to be.

As a concrete indication of the puzzles this raises, consider the following passage from Kitcher (1993: 92):

> Our definitions should be subject to criteria of adequacy in that they need to make it clear how achieving a progressive sequence ... would realize (or realize more closely) our goals. The account that follows will presuppose that there are goals for the project of inquiry that all people share – or ought to share ... the goals in question are impersonal ... We need a specification of impersonal goals for science, goals that can ultimately be defended as worthy of universal endorsement.

Kitcher (1993) doesn't explain why we 'need' impersonal aims. ('Goals' and 'aims' are synonymous.) He doesn't explain how such putative aims arise or are grounded, although he later purports to tell us what they are. He also suggests that 'we ought to share' these aims. But what kind of normativity is involved in that claim? The passage is assertive in style but cryptic in substance.[26] It leaves us with the following questions, among others. Are the impersonal goals already 'out there'? If so, where? How do we identify them? Or is Kitcher just *proposing* goals that are worthy of universal endorsement? If so, on what grounds would they be so worthy? Impersonal values? If so, where do those arise? And how do we identify them? If not, then what other grounds are there?

Answering 'what is it to be an aim of science?' might help us to resolve such concerns. As I will now show, however, the existing answers to this question are either problematic or underdeveloped. Moreover, even the most promising accounts of scientific aims do not cohere with the most prominent positions on (first-order) goodness makers or with the methods philosophers of science use to argue for those positions.

I take coherence of this kind to be important for reasons of charity. Thus, any answer to 'what is it to be an aim of science?' entailing that the work covered in Section 1 is misguided or unfounded is undesirable, ceteris paribus. If it should transpire that the only independently plausible answers have such consequences, however, then the bitter pill should be swallowed and more radical alternative views on second-order goodness makers – such as the account in Section 3 – should also be considered.

2.2 Existing Accounts of 'The Aim(s) of Science'

Resnik (1993) critically examines four potential accounts of 'the aim(s) of science'. These are as follows:

(1) Shared goals
(2) Corporate goals
(3) Normative ideals
(4) Characteristics

I will consider each of these approaches and summarise their defects. I will then consider whether an account of collective goals – based on more recent work on collective intentionality – has better prospects. I will conclude the survey of existing accounts by considering Bird's (2022) proposal, which appeals to the aim of belief and proper functions.

---

[26] Kitcher (2015) now distances himself from such earlier proclamations.

2.2.1 Aims As Shared Goals

The first account appeals to an intentionally sought goal of all or most scientists. However, authors on scientific progress appear not to have had this view in mind – and would have been methodologically remiss if they did – because they have done minimal empirical spadework to support their views of what (first-order) goodness makers are.[27] Indeed, as Resnik (1993: 226) pithily demands of anyone who prefers this construal: 'show me that set of aims and prove to me that most scientists have those aims'.

The view that scientific aims are shared goals is also in direct tension with several existing accounts of first-order scientific goodness makers. Here are two arguments to this effect.

First, scientists lack philosophically sophisticated goals. To illustrate, imagine one wanted to argue for Bird's (2007) epistemic view of progress by showing that the aim of science is to attain knowledge. Do scientists aim to attain knowledge in the same sense that Bird, who endorses Williamson (2000) on knowledge, intends it? No. They lack that concept of knowledge.[28] Do scientists instead have the goal of knowledge despite having a minimal under-standing of knowledge? Do they want to achieve the goal without having a good idea of what that goal is (or what it would look like to have reached it)? The problem is that multiple scientists proclaiming to be after 'knowledge' might each have significantly different goals from an epistemological perspective. Moreover, it would be insufficient to claim that scientists seek knowledge because they're aware of the consequences of having it. Their goal would then be those consequences; knowledge would merely be a means to an end.

Second, more trenchantly, if the aims can change as the community changes, then what counts as cognitive scientific progress can change too. However, as we saw in Section 1, most existing authors state that there is one atemporal, immutable, cognitive goodness maker (or set of goodness makers). Adequate support for such views would require substantial longitudinal evidence about past science, in addition to evidence about present science, if shared goals determine what counts as progressive. Yet authors on scientific progress do not provide anything approaching this. They typically use isolated case studies and thought experiments, which do not refer to what the scientists therein are trying to achieve.

---

[27] This has only recently changed, with the work of Mizrahi, as we saw in the survey. Most engaged in the debate do not take such work to be relevant in telling for or against their views on progress.

[28] This claim depends on an internalist view of concepts, which I cannot defend here. Let me emphasise, however, that I am not committed to the idea that there is no overlap between the extension of 'knowledge' for scientists (as a whole) and 'knowledge' for, say, knowledge-first epistemologists.

2.2.2 Aims As Corporate Goals

The second option compares science to a corporation. Corporations might be said to have aims insofar as they have hierarchical structures designed to achieve specific goals. Such structures can operate to achieve those goals irrespective of what those within the organisation personally aim at. Thus, McDonald's can be said to aim at dominating the world market for fast food, although almost no one working in any of its restaurants has that aim, and few of its senior managers have that aim. This view is also less susceptible to the concern that the goals won't be stable over time, especially if a baptismal event set them. For instance, a corporation may have goals stated in its charter. Its raison d'être may be to achieve them.

But does science have aims in the same way that corporations do? One reason for thinking not, which Resnik (1993: 227) focuses on, is that science doesn't have a hierarchy of the kind that corporations have: 'a system in place that assigns its members a certain rank which confers on them the power to make certain decisions and delegate authority'. Nor was there a baptismal event during which aims for science were chosen.

A possible riposte is that corporations can be less autocratic than Resnik admitted and may introduce goals over time. To the extent that this is a retreat to the shared goals approach, it raises similar problems. Nonetheless, perhaps there is room to allow that science has implicit aims, because it has been constantly redesigned to perform (or more reliably perform) specific functions. This suggestion foreshadows some of the discussion that follows. Here is an initial response to it.

Science involves highly diverse activities and starkly different approaches across disciplines and sub-disciplines. For example, a molecular biologist I once interviewed declared that a condensed matter physicist's theoretical model of a membrane was 'oversimplification gone mad! A perfectly spherical horse!' (Rowbottom 2011b). So even if science reliably produces particular 'goods', it is unclear that this is the result of intentional design choices reflecting implicit aims for the enterprise. Design choices *have* affected how elements of science work; and hence at a smaller scale – at the level of research groups, and perhaps sub-disciplines – aims of this kind operate. However, the presence of low-level, corporate-style aims *within* science does not suggest the existence of any aim(s) *of* science.

2.2.3 Aims As Normative Ideals

The third distinct option is to construe scientific aims as goals that scientists should ideally pursue. Resnik (1993: 228) argues that this approach is double-edged: it frees us from the need to empirically investigate science at the cost of leaving us with no clear way to settle the question of what scientific aims are, in the face of remarkable dissensus among philosophers of science, except via intuitions. This argument seems too quick, at first sight. Why not classify goals in current or historical science as closer or further away from ideal? The answer is that this doesn't provide a straightforward way to settle the disputes by appeal to actual science. One can only determine how far away an actual situation is from an ideal state of affairs when one has something about the latter in mind. The most obvious way to address this difficulty is to derive the ideal state from some features of current and historical states of science. For instance, one could consider collective intentionality across some such states. But this would involve hybridising the view.

A further problem with this approach is that philosophers of science tend to think of good and bad science in consequentialist terms. They hold that if setting up science in a unique way maximises the objective probability of it producing 'good' outputs – approximately true or empirically adequate theories, say – then that is how it should be set up. However, it is dubious that there is any specific shared or common aim that scientists ought to have from such a perspective. Social epistemologists of science, such as Kitcher (1990; 1993), Longino (2001), and Muldoon (2013), tend to argue that diversity among agents is valuable. Moreover, Rowbottom (2011a: 124) argues not only that science can profit from scientists with irrational aims, such as preserving their preferred theories come what may, but also that 'ideal science may be realizable in more than one way'. Hence, associating ideal science with any unique set of aims on the part of its practitioners might be a fundamental error.

2.2.4 Aims As Characteristics (Or As Constitutive)

The final option is to explicate aims of science as characteristics of science. Van Fraassen (1998: 213) adopts this approach: '[s]cientific realism and constructive empiricism are ... not epistemologies but *views of what science is*. Both views characterize science as an activity with an aim'. One minor concern with this approach is that we cannot specify how to demarcate science from other enterprises, such as pseudoscience, except in highly abstract terms.[29] As Resnik (1993: 299) points out, however, it might nonetheless be possible to characterise science in a somewhat weaker sense, by understanding 'aims of science' to refer to 'a set of common characteristics [that] apply to most of the things we call scientific'.

---

[29] Consider, for example, Hoyningen-Huene's (2013) recent attempt to demarcate science by appeal to systematicity. Hoyningen-Huene (2013: 169) admits: 'the statement of a difference in the degree of systematicity between two areas is extremely abstract and therefore "thin" or, to put it in more unfriendly terms, very vague. Not much is put on the table by such a statement, and correspondingly, it will be difficult to refute it'. Nevertheless, Oreskes (2019) makes a good attempt at refuting it.

A deeper problem is that many characteristics of actual science, so construed, have nothing to do with science's 'aims' in an appreciable metaphorical sense. Consider banal but accurate statements such as 'Science involves mathematics,' 'Science takes time,' and even 'Science produces many documents containing the letter *e*.' Do these specify aims of science? The advocate of the view of aims as characteristics will presumably answer in the negative and explain that the aims of science are a proper subset of the characteristics of science. But then we are not much closer to specifying what aims of science are. The question is now: what picks out any given characteristic of science as an aim rather than something else?

Van Fraassen (1994; 1998) answers by appealing to success in science as such. This suggestion is problematic partly for reasons adduced by Rowbottom (2014), with reference to van Fraassen's examples of chess and war. Chess is a two-player game where the criterion for success is stipulated. Science isn't like that. Wars – which are better cases to consider insofar as they involve many different people, groups, and activities – have concluded repeatedly. There are many examples of end-states that we can consider. Science isn't like that either. Furthermore, we look to the aims of those responsible for orchestrating wars, to determine who won in any individual case, which we should avoid doing according to the 'aims as characteristics' approach (lest it collapse into a corporate-style aims account).

2.2.5 Collective Intentionality and Goals

A significant body of work on collective intentionality has appeared since Resnik (1993), as detailed, for example, by Niiniluoto (2020) and Schweikard and Schmid (2020). So let us now see if this provides a viable alternative to shared or corporate goals accounts.

Perhaps the most influential approach is Tuomela's (2007: 145), where 'groups can but need not be taken as (singular) entities, and . . . are agents and persons only in a metaphorical sense', and where *we-intentions* underpin the legitimate ascription of group attitudes. Miller and Tuomela (2014) explore how to construe collective goals in this vein. I will consider only the 'weakest' of the accounts they propose, or the easiest to satisfy, because this is the most defensible as an account of scientific aims.

On said account, *we-intentions* involve desires and beliefs, but not intentions: '[a] collective goal based on shared "we-wants" (or "we-goal") . . . is the weakest notion, in which the central connecting "social glue" is the participants' mutual belief. It does not even require that the members intend to achieve the goal. Wants are intentional in the "aboutness" sense, but one can want something without intending to achieve it' (Miller and Tuomela 2014: 43).

A minor worry about this proposal is that no requirement for the 'we-wants' to be achievable – or, at least, not to be believed to be unachievable – is specified. Hence, a collective goal in this sense could be 'utopian', as Laudan (1984) put it. But it would be easy to alter the account to preclude such goals, so we need not dwell on this.

A more pressing problem is that the 'social glue' element involves individuals having beliefs *about what others believe*. As Miller and Tuomela (2014: 44) explain, instead of considering group goals in a 'we-mode' sense where the requirements are stringent:[30]

> One may speak of group goals in a weaker, I-mode sense: Group G has the goal P to the extent its members share the 'we-goal' to achieve P. A we-goal in the I-mode sense is defined in a stylized way as follows: a member has the we-goal P if and only if she has goal P, and believes that the others in the group (or most of them) have P and that this is mutually believed in the group. When the we-goal P is shared by all or most of the members of the group, we can speak of a group goal: to the extent P is shared, the group has the goal P (in the I-mode).

Hence, this account builds upon a shared goals approach and introduces additional belief-based requirements. Specifically, for P to be a we-goal (in the I-mode sense) most scientists must not only have the goal P (as in the shared goals account) but must also believe: (1) that most other scientists share the goal P; and (2) that most other scientists believe that 'most other scientists share goal P'.

Are these extra requirements satisfied in the case of science? It is doubtful. Many scientists would be reticent to ascribe general goals such as 'finding knowledge' or 'finding as much of the truth as possible' to most other scientists. They would be more reticent still to ascribe beliefs to most other scientists *about* the goals of most other scientists. Why? Because scientists are often aware of the striking differences, remarked on previously, between different scientific disciplines and sub-disciplines.[31]

---

[30] I do not consider the we-mode variant or watered-down variants because these are much harder to defend as pertinent to goals of science (rather than of research groups, or such, therein). The we-mode variant is characterised as follows:

> Goal P of an agent x is in the we-mode relative to group g (or x has P in the we-mode) if and only if (i) x is functioning qua member of g, (ii) goal P has been collectively accepted or at least is based on what the members collectively accept and, consequently, (iii) it satisfies condition (c) [which concerns P having collective content], (iv) it serves as a group reason for the members, (v) x and the other members are collectively committed to satisfying P, (vi) P satisfies the collectivity condition, and (vii) x intends or wants to participate in the satisfaction of P for g. (Miller and Tuomela 2014: 39)

[31] Such differences have been discussed at length by Galison (1997) and Keller (2002); scientists' awareness of them is further illustrated by many of the interview excerpts in Rowbottom (2011b).

As this is an empirical hypothesis, it is open to refutation. Yet philosophers working on scientific progress have not adduced any empirical
evidence showing the existence of collective aims (even if there are now
slivers of evidence about shared goals). I take the burden to be on them to
provide it.

My other concerns about the collective view mirror those I have about the
shared goals approach. In essence, I doubt that many authors on progress
would endorse this view when the current lack of evidence about collective
aims would undercut their claims about what scientific progress consists in
(and make their earlier work appear highly speculative as a result). Several of
these authors have argued for decades that science has a specific aim, without
adducing any evidence concerning what scientists believe about their own
aims or the aims of others. They have shown minimal sensitivity to the notion
that aims might change over time, in advancing their monistic views of
goodness makers. The most natural explanation is that they do not think
their claims are empirically refutable. If they were, indeed, then why would
an inquiry into goodness makers be especially *philosophically* interesting? It
would primarily be a matter for social scientists to investigate and report on.
For the sake of completeness, however, I will close by considering if there is
any way for such philosophers to bite the bullet and posit something in the
region of collective goals.

Miller and Tuomela's (2014) account of collective goals – like the corporate
goal account – licenses talk of many organisations having aims. But as we
have seen, there are significant disanalogies between such organisations and
science. Most notably, there is no identifiable *collective origin* for putative
general goals of science, such as an explicit collective decision to pursue those
goals.[32] So if a collective account of science's aims is to become tenable, then
collective commitment to global goals – like increasing verisimilitude – must
emerge by a different route, such as via shared practices (which *do* involve
collective local goals).

One possibility is that participation in local activities in science creates a tacit
acceptance of a specific global purpose of science. Alternatively, it might be
possible to consider how *groups of groups* – astronomers and botanists, for
example – can have collective aims that arise from their choices to affiliate to

---

[32] Miller and Tuomela (2014: 38) note the importance of such origins, although they don't make
having a collective origin a necessary condition of having a collective goal:

> 'Besides having collective content, x's collective goal often has a collective origin that
> can be revealed by tracking the "history" of the collective goal. The collective goal
> might have been the result of, and engendered by, an action of collective decision or
> acceptance.'

some extent and to share an ethos.[33] For instance, The Royal Society was originally called 'The Royal Society of London for Improving Natural Knowledge'. Thus, one might take it to have the goal of improving natural knowledge. But one might find other bodies that serve different goals according to their mission statements. So how to aggregate? Moreover, science predates the existence of such societies and bodies (according to most historians and philosophers of science). Even from a highly optimistic perspective, we only have a promissory note for such an account.

I will close this section, however, by outlining the most promising approach I envisage and explaining why I am not sanguine about its prospects. The underlying idea is to look for practices common across all the sciences, which were introduced to perform cognitive functions. One practice that stands out is peer review. Might one not argue that this promotes increases in truthlikeness, knowledge, or some such? Might one not argue that it was introduced – and has spread so widely – because it performs this function?

There are two difficulties with such lines of argument. First, peer review only selects for what is considered 'good' by the reviewers in the context; and even assuming that reviewers are invariably after some lofty goal such as promoting increases in truthlikeness, how exactly they pursue that, and the extent to which they succeed, varies considerably. But perhaps this could be put to one side. Perhaps whether peer review promotes what it was intended to promote is irrelevant. Perhaps only the intent matters.

But second, according to historians of science, peer review is used for different reasons across contexts; there has been no single intent in introducing it. As Moxham and Fyfe (2018: 863) explain:

> Refereeing emerged as part of the social practices associated with arranging the meetings and publications of gentlemanly learned societies in the late eighteenth and nineteenth centuries. Such societies had particular needs for processes that, at various times, could create collective editorial responsibility, protect institutional finances, and guard the award of

---

[33] This is similar to (but weaker than) the approach suggested by Niiniluoto (2020), who writes that: 'Principle n is a norm of science if and only if n is jointly accepted by the scientific community'. He continues:

> This means that all (or most) members of the scientific community follow n in their behavior and expect that all (or most of) the other members follow n. A similar analysis can be given of the values of science: they are not simply virtues or attributes of individual scientists but institutional commitments ... universities and research institutions are value-based organizations whose ethos includes goals like truth, knowledge, critical attitude, creativity.

The outstanding issue is how this bears on science itself, rather than any specific organisations that have scientists as members.

prestige. The mismatch between that context and the world of modern, professional, international science, helps to explain some of the accusations now being levelled against peer review as not being 'fit for purpose'.

They go on to conclude that:

> [T]he relative durability of refereeing as a practice should not be mistaken for simple continuity of purpose or meaning. What it was meant to accomplish, whom it was intended to benefit, and the perception of its virtues and defects varied considerably with time and place … the nature and purpose of refereeing, and of peer review, vary importantly with context. (Moxham and Fyfe 2018: 888–9)

Note the mention of 'nature', in addition to 'purpose'; review procedures vary in interesting ways – blinding procedures, number of reviewers, visibility of review to the community, editorial role, and so forth – that bear on what it achieves. The more closely one looks at such practices and their histories, the more one becomes aware of significant differences between them.

2.2.6 More on Functions: Against Bird's New Approach

Recently, Bird (2022) has suggested that functions underpin aims. I should add something on this, although I must be brief. He writes:

> In my view, belief has an aim because [it] is governed by norms of correctness, and these norms arise from the fact that our belief-producing (i.e. cognitive) systems have a *function*. A correct belief is one produced by a properly functioning cognitive system. And a belief produced by a properly functioning system is knowledge. So beliefs aim at knowledge. In order to draw the same conclusion for science … we should understand institutions such as science as having functions. Science, in particular, has a cognitive function. (Bird 2022: 15)

I have several objections to this approach. First, it pushes the debate back into fundamental metaphysics and epistemology (and related areas such as philosophy of mind), increasing its complexity and resultant dissensus, while moving it further away from science studies. For instance, there are numerous ways to construe beliefs, and there is a vast literature on the ethics of belief (which covers what it is to be an aim of belief, as well as what such aims are). For example, one might be a dispositionalist about belief (following Schwitzgebel 2002) and consider the function of belief in an evolutionary light. One might take beliefs to be 'correct' in this picture when they foster dispositions that promote survival. One might end up with a pragmatic epistemology, following Stich (1993), which would fit well with a functionalist account of progress. But should we be arguing about this kind of thing to determine what the aim of science is (or whether there is one), if we grant that science performs cognitive functions?[34]

Second, on a related note, the approach is limited in taking goodness bearers only to be beliefs (of individuals or groups). The scope of 'cognitive' is extremely narrow on this view, and the motivation comes from only one strand of contemporary epistemology (namely, the knowledge-first approach). As we saw in Section 1, however, science develops in several ways that appear, prima facie, to be cognitively relevant. Know-how increases. Understanding increases. Problem-solving ability increases. What's published – and what can *possibly* be, although is not actually, believed – also changes. Again, armchair thought about limited aspects of humans – their belief systems – does not seem to be an appropriate substitute for the study of actual science (even if that sometimes proceeds by thought experiments). This is especially so when even the psychological reality of beliefs is contested.

Third, it is incorrect that 'a belief produced by a properly functioning [cognitive] system is knowledge'. Whether or not one's (properly/rationally/ justifiably formed) beliefs amount to knowledge is determined by one's environment. For example, even a paragon of rationality will generate many false beliefs if placed in a computer simulation of a pre-industrial society where Gettier scenarios abound. Bird might rejoin that the proper function is revealed by the way the system performs in the *right* kind of environment, but then we are left to ask why science should be formed to work in such a 'right kind of environment', rather than the actual environment.

Fourth, and finally, 'X has the function of Y' does not entail, or even appear to significantly raise the probability of, 'X has the aim of Y'. Consider the heart as a case in point. Functional discourse concerning this goes back to Galen, at least, and many would be content, in the modern day, to allow that it (roughly) performs the function of pumping blood around the body. But does the heart have the *aim* of pumping blood around the body? No. To asseverate otherwise is to make a category mistake. A cardiologist would not say 'Your heart is now better achieving its aim' after performing a bypass operation, although she might say 'Your heart is functioning better.'

---

[34] Bird could retort that I similarly proclaim the significance of meta-normative questions for work on (first-order) goodness makers. There is, however, an underlying difference. The view I develop has a strong empirical component, insofar as it concerns what science can achieve (with varying degrees of aleatory probability). Bird's approach is more from the armchair. His findings for science presumably go for *any* form of inquiry (performing cognitive functions), from art history to chess studies. Knowledge is the aim. This goes against my instincts as an empiricist. My ultimate concern is that when we work at such a level of abstraction, we idealise science so much that we lose sight of its real – warts and all – character.

Similarly, institutions with aims may perform functions that are only indirectly related to those aims. So even granting that X has an aim, it does not follow from the fact that Y is a function of X that Y is an aim of X. Consider banks. Many perform the function of lending money. But banks aim to make profits. If lending ceases to generate profits, they can legitimately cease doing it. They can instead focus on other financial means to achieve their core aim, such as investing deposited funds.

A natural counterproposal would be that functions operate *directly* as second-order goodness makers (and that aims may be removed from the picture). But this is not much of an improvement because a significant gap remains between 'X has the function of Y' and 'X makes progress when it achieves Y or gets closer to achieving Y'. Does a heart make *progress* when it pumps blood around the body more efficiently than before, or when it restarts after unexpectedly stopping? No. Even the metaphorical use of 'progress' is clumsy in this context.

2.2.7 Hybrid Accounts of Aims, Appeals to Origins, and an Intermediate Conclusion

I will make two further points before offering an intermediate conclusion and explaining how I will proceed.

First, hybrid accounts of scientific aims, which appeal to more than one of the prior accounts, are possible. However, these are relatively unexplored. Perhaps the most promising option would be to consider collective or corporate aims (or some appropriate admixture) characteristic of science. But taking such a route would be even more challenging than taking a collective or corporate aims approach in isolation and would not obviously serve to rectify any of the problems with those views.

Second, one cannot avoid appeal to aims by adopting an 'originary' account, following Kitcher (2015: 478), according to which 'progress is measured in terms of the distance from the starting point', and rejecting a 'teleological' account where 'there is a goal state and when progress consists in increased proximity to that state' (478). To take an originary approach is *not* to specify what second-order goodness makers are; it is simply to reject some views of aims, such as the normative ideals option. One way to see this is as follows. It can be a legitimate aim to increase something, whatever one's starting state happens to be, without referencing a 'goal state'. For instance, I can aim to increase my fitness by running, just as scientists might aim to find theories closer to the truth than those they start with. Thus, Kitcher's use of 'teleological' is misleading; avoiding such a 'teleological' view is compatible with holding that the aims of science are second-order goodness makers.

This brings me to my intermediate conclusion. Given the inadequacy of the
accounts of aims as normative ideals or as characteristics, the authors on
progress covered in the survey face a dilemma.[35] On the one hand: adopt
a collective or corporate account of aims (trusting that the problems with
these are remediable), admit that there is minimal evidence about what first-
order goodness makers are, renounce your previous appeals to thought experiments or isolated case studies, and recognise that serious historical and social
scientific studies of science are the way to proceed in studying scientific
progress. On the other hand: recognise that there is no satisfactory account of
the aims of science, entertain the possibility that there are no such aims, and
explore the consequences.

I will go down the second route in what follows. In the remainder of this
section, I will provide some necessary conditions for something to count as
a second-order goodness maker (qua 'aim of science'). In the next section, I will
apply these conditions to develop my own meta-normative account of scientific
progress.

2.3 Necessary Conditions for a Satisfactory Account of 'the Aims of Science'

Although the existing accounts of 'the aims of science' are problematic, they
highlight important considerations for deciphering the metaphor. They suggest
that any adequate account will hold, inter alia, that:

(i) The aims of science involve or relate to the aims of some scientists.
(ii) The aims of science are characteristic of science.

The first requirement arises because the metaphor must derive its meaning, if
apt, from a consideration of the aims of people (or groups thereof) – genuine aim
holders – in doing science. Similarly, if someone says 'the aim of jogging is to
keep fit', we take them to be making a claim about why some people do, or some
people might reasonably, jog. This is compatible with jogging having
a baptismal aim.

The second requirement is appropriate because authors on the aims of science
take themselves to be determining something significant about the real activity,
science, as distinct from other activities. This constraint is relatively weak, so
should be uncontroversial.

---

[35] If one takes Bird's appeal to the aim of belief to be promising, then one might view this as
a trilemma instead. The third option would be to stop looking at science so much and start doing
more fundamental philosophy (because that's where the question would be settled, if anywhere).
See Section 2.2.6.

2.3.1 Rational Aims As Characteristics

How might we build upon this foundation? One promising way is to require that:

(iii) For any X to be an aim of science, X must be a rational aim.

Popper (1983: 132) flirts with this idea:

> [S]cience itself . . . has no aims . . . Yet when we speak of science, we do seem
> to feel . . . that there is something characteristic of scientific activity, and since
> scientific activity looks pretty much like a rational activity, and since
> a rational activity must have some aim, the attempt to describe the aim of
> science may not be entirely futile.

To motivate (iii) further, consider how it coheres with several philosophers' views about scientific aims and scientific progress. First, recall Laudan's insistence that the aims of science should not be utopian; if they were utopian, they would be irrational, but science is a rational activity for Laudan. Second, remember Kitcher's requirement that scientific aims should be worthy of universal endorsement. They would fail to be worthy of such endorsement if they were irrational. Hence, rationality is required. Third, no contemporary author on progress argues that making progress involves increasing X while accepting that increases in X are either impossible or a matter of mere luck. On the most influential accounts of progress, on the contrary, scientists can reliably increase X, qua scientists. For example, Bird thinks that scientists can reliably increase scientific knowledge – and that properly functioning cognitive systems reliably produce knowledge – whereas Kuhn thinks that normal scientists can reliably solve problems.

We can make the rationality requirement for aims, (iii), more precise by considering *hypothetically rational aims in doing science*.[36] The rationality at issue here is objective. It concerns what doing the activity can achieve, rather than what a person might expect to achieve in doing it. For example, one might say:

> X is a hypothetically rational aim in doing science if and only if doing
> science is a reliable means of achieving X.

Then by explicating reliability in terms of aleatory probability – as is standard in reliability engineering, where reliability is unity minus probability of failure – one might say:

> X is a hypothetically rational aim in doing science if and only if doing
> science has an aleatory probability of achieving X above a threshold value
> of n.

---

[36] The underlying idea is that for X to be an actual rational aim, X must be a hypothetically rational aim. So understanding what it takes to be hypothetically rational helps to understand what it takes to be rational.

This leaves the delicate matter of choosing an appropriate value for n.[37] However, most would agree on a range of values falling below and above the threshold – for example, $[0,0.5]$ and $[0.75,1]$ – so I don't consider this worrisome. I hold this embryonic account of hypothetically rational aims to be inadequate for two different interrelated reasons.

First, it is too broad. For instance, 'changing our opinions about some propositions' is a hypothetically rational aim on this definition. It is cognitive. But it can be achieved easily by other means, such as reading a newspaper, surfing the Internet, or having a mundane conversation. The underlying problem is that doing this isn't characteristic of science as distinct from other forms of activity, including pseudoscientific activities such as reading tarot cards. Requirement (ii) isn't satisfied by the initial account, because it doesn't refer to what doing science is especially likely to achieve.

Second, it is dubious that n should have a fixed, contextually invariant, value. To see this, imagine that doing science significantly increases the aleatory probability of achieving a possible goal, such as finding the true theory of everything, although achievement thereof remains extremely unlikely. Or imagine that doing science is the *only* possible way to raise the aleatory probability of achieving that goal above zero. Doing it with that goal in mind might then be rational. Similarly, the (objectively rational) aim of buying a lottery ticket might be to win the lottery although doing so only provides a miniscule chance of winning. Context is relevant in making such a judgement. If one had a way to guarantee winning the lottery that was as easy as buying a ticket, such as fixing the lottery, then buying the ticket wouldn't be rational (provided other goals, such as obeying the law, weren't relevant).

Both issues could be considered in much greater depth, and reasonable differences of opinion on how to resolve them are possible. For example, some might endeavour to defend the idea that n should have a fixed value. However, my aim is to emphasise the utility of the notion of hypothetically rational aims in doing science. I do not have the luxury of determining which analysis of this concept is best. I can only identify some key analyses and explain why I find one of these promising.

The two issues identified previously can be addressed by revising the initial account as follows:

> X is a hypothetically rational aim in doing science if and only if doing science is the most reliable means of achieving X.

---

[37] Space prevents me from discussing which aleatory interpretation of probability is suitable in this context. A long-run propensity account – see Gillies (2000: ch. 7) and Rowbottom (2015b: ch. 8) – is my preferred option.

Thus, when (degree of) reliability is defined in terms of aleatory probability, it follows that:

> X is a hypothetically rational aim in doing science if and only if doing science raises the aleatory probability of achieving X more than doing any other possible activity does.[38]

This is the basic statement of the account of hypothetically rational scientific aims that I propose. For cognitive aims, it is more precisely:

> X is a hypothetically rational *cognitive* aim in doing science if and only if doing science raises the aleatory probability of achieving X more than doing any other possible activity does and X is of a *cognitive* variety.

I will explain how this account might be refined in the next section. Beforehand, however, I should like to emphasise that it is compatible with ranking some hypothetical aims – and, naturally, some *actual aims of scientists* at any point in time – as more significant than others.[39]

Consider the way that philosophers such as Popper (1983) and van Fraassen (1980) write of *the* aim of science, and how other philosophers, such as Bird (2007) and Dellsén (2016), adopt monistic or narrow views of progress. How can we adjudicate such claims? Perhaps with recourse to the relative extents to which doing science raises the aleatory probability of those (hypothetical) ends being achieved. This isn't to presume that empirical adequacy wins out, even granting that generating empirically adequate theories – or more accurately, clusters of laws, hypotheses, and models – is easier than generating true equivalents. Grant that the aleatory probability of science achieving X is greater than the probability of science achieving Y. It doesn't follow that doing science (rather than something else) *increases* the probability of achieving X more than it *increases* the probability of achieving Y. Nor does it follow that the probabilities of achieving X and of achieving Y are independent.

My intent here is not to determine how to rank aims – actual or hypothetical – in terms of centrality. It is merely to point out that reliability-related considerations are potentially relevant to how such ranking should be done.[40]

---

[38] Possibility in practice is intended; an objectively rational thing to do, like a morally obligated one, must be possible to do.

[39] As Niiniluoto (2019) puts it, 'there is no reason to assume that the goal of science is one-dimensional', and as Rowbottom (2019: ch. 1) emphasises, one might allow that there are many goals and maintain that they can be ranked in terms of centrality.

[40] See also Okasha (2011) on the ranking of theoretical virtues (and Kuhn's 'no algorithm' thesis on this); much of this discussion goes, mutatis mutandis, for aims.

their speakers presuppose that second-order goodness makers are objective or intersubjectively privileged standards. Claims about scientific progress (or regress) having occurred are also false whenever their utterers presuppose that second-order goodness makers are objective or intersubjectively privileged standards.

4. Claims of the form 'Scientific progress consists in X' are true when their speakers assume the operant cognitive standards to be imposed by individuals or groups, provided that: X is a hypothetically rational aim in doing science; and the pertinent individuals or groups value X. Moreover, claims about the occurrence of scientific progress (or regress) may be true when they are indexed to an X satisfying the aforementioned conditions.

5. Philosophers of science would typically be better off not using 'progress'. Talk of imposed standards or values can achieve the same ends in a more pellucid fashion.

I will also defend some other, non-core, claims. For instance, I will argue that the first two theses explain why there is so much dissensus about what first-order goodness makers are. And I will suggest that the objectification of standards performs the function of bestowing authority on (various elements of) science. Furthermore, I will explain that I take my principal findings to go for philosophy too.

Unfortunately, I do not have the luxury of engaging in depth with the extensive literature on meta-normativity within the confines of this Element.[42] Thus, I cannot seriously consider all the key alternatives to my position – for example, most notably, cognitive analogues to moral naturalism and moral constructivism – while arguing for it. I do, however, introduce some arguments that don't have meta-ethical parallels. Moreover, my position serves as a foil, to prompt further work on this neglected issue, even if I cannot defend it as well as I would like due to a lack of space.

3.1 What I Do Not Deny: Local Cognitive Aims and Progress

Before I explicate and defend these core theses, I should highlight two claims that I accept.

First, there is an objective sense in which 'progress' is possible once standards are accepted or imposed. Or, as I'd rather put it, there is an objective fact about the extent to which changes result in movement towards or away from a specified state. Mackie (1977: 26) put it so:

---

[42] I am also not able to cover several pertinent meta-normative issues. See Baker (2018) for a useful overview of these. The distinctions between constitutive and regulative norms and between procedural and substantive norms are of special relevance to the scientific progress debate.

> Given any sufficiently determinate standards, it will be an objective issue,
> a matter of truth and falsehood, how well any particular specimen measures
> up to those standards. Comparative judgements in particular will be capable
> of truth and falsehood: it will be a factual question whether this sheepdog has
> performed better than that one.

Second, as explained in Section 2, I accept that individual scientists
and groups thereof have cognitive aims (which are often rational) and
that progress can occur from their perspectives due to specific scientific
changes. Thus, I have no objection to the claim that 'Gravitational wave
physicists made significant progress in detecting a gravitational wave',
provided that a gravitational wave was indeed detected. Finding such
waves was a relatively local, specific, aim of a community in doing
a whole host of things, including building and developing the LIGO
installations at Hanford and Livingston. Such aims, which are many
and varied, change regularly. Indeed, the aims of the community of
gravitational wave physicists have changed because gravitational waves
are now thought to have been detected.

3.2 On the Non-existence of Objective Standards [Thesis 1]

My position on cognitive standards in science is inspired by Mackie's (1977)
position on ethical standards.[43] In particular, I deny the existence of any
objective standards pertaining to cognitive progress in science (or that
objective second-order goodness makers exist).

Partly, I deny this because the standard route to claiming that there are such
standards, via appeal to objective aims of science, fails. There are no such objective
aims, as argued in Section 2. There is no objectively ideal state of science, as
distinct from what people consider ideal due to their personal or group values.
There was also no 'baptismal moment' where cognitive goals were *intersubjectively* chosen for the entire enterprise of science. And even if there had been, that
might just reflect people's aims in setting up the activity. This would be compatible
with people later electing to do it with different aims, even if they were aware of –
and took themselves with false consciousness to be pursuing – the baptismal ends.

Partly, I deny this because the very idea of such standards is odd. First, how
could such objective standards – governing a human activity – arise without the
presence of some higher authority, such as God? Second, how could we come to

---

[43] However, I am sympathetic to the non-standard interpretations of Mackie's position detailed by
Moberger (2017) and Berker (2019). I do not think that what normally passes for error theory in
meta-ethics reflects what Mackie meant by 'error theory'. I call my position 'quasi-error
theoretic' only because it is distinct from the contemporary meta-ethical view normally picked
out by 'error theoretic'.

know of such standards? Wouldn't we need a mysterious faculty of intuition, granting us awareness thereof?[44] In essence, these are the two strands of Mackie's (1977: 38) original argument from queerness:

> If there were objective values, then they would be entities or qualities or relations of a very strange sort, utterly different from anything else in the universe. Correspondingly, if we were aware of them, it would have to be by some special faculty of moral perception or intuition, utterly different from our ordinary ways of knowing everything else.[45]

Since this argument has been so widely discussed in the moral context, I will not dwell on it.[46] Instead, I will propose a rather different argument concerning scientific cognitive standards. I call this *the argument from the lack of explanatory significance*. The fundamental idea is that positing objective standards does no explanatory work in the scientific case.

In the moral case, such standards potentially explain the presence of normative force, and hence our feeling that there *really* is (normally) something wrong with performing an abortion in the third trimester. Yet it is dubious, for instance, that scientists who uncritically accept dominant flawed theories and try to refine minor aspects of these are *really* doing anything wrong. Nonetheless, said scientists are not contributing to progress on the most popular contemporary accounts thereof.

Furthermore, there is no compunction to do science (or normative force from without). I don't think that a society which abandoned science altogether, despite having the capacity to do it without any severe cost, would be remiss. A strong argument would be needed to make us conclude that doing science – or doing science in any particular way – is obligatory (rather than supererogatory).

A third and final argument against the existence of objective standards is the argument from relativity, which Mackie (1977: 36) presents as follows: 'radical differences between first order moral judgements make it difficult to treat those judgements as apprehensions of objective truths'. And the dissensus involving first-order cognitive goodness makers in science – and hence in judgements about when progress has occurred, what has brought it about, and what actions would promote progress in the future – is perhaps even greater than the equivalent in the moral case. Moreover, these views are advanced by alleged experts on scientific progress, rather than 'the folk'.

---

[44] As intimated in the introduction to this section, there are potential constructivist and naturalist answers to these two questions. See Bagnoli (2021) and Lutz and Lenman (2018). However, these face challenges. For instance, Shafer-Landau (2003) argues that constructivism relies on arbitrary standards or collapses into realism.

[45] Mackie writes of values rather than standards in this passage, although, as we will later see, he referred to 'standards' on other occasions.

[46] See, for instance, Tännsjö (2015) and Moberger (2018).

3.3 On the Falsity of Discourse Presuming the Existence of Objective Standards or Privileged Intersubjective Standards [Thesis 3]

If objective standards pertaining to cognitive scientific changes do not exist, then what follows concerning claims that presume their existence? Like Mackie on the standard reading, I hold that such claims are false.[47] The argument is straightforward. Consider a statement like 'Scientific progress occurs precisely when scientific knowledge increases' and grant that such a statement might have different proper interpretations in different contexts. If the proposition expressed by the statement in context (or the proper interpretation of the statement) implies 'The objective aim of science is knowledge' then that proposition is false. (If *p* entails *q* and *q* is false, then *p* is also false.) Hence, the statement is false in the precise sense that its proper interpretation is false. But naturally, there may also be ways to misinterpret it such that it appears true. The joke sign that often appears in pubs, 'Free beer tomorrow', relies on such ambiguity.

Unlike Mackie, however, I am not concerned only with 'ordinary . . . thought and language', but also with how philosophers have discussed progress. (I do *not* make a general claim about the falsity of statements of the form 'Scientific progress occurs precisely when *p*'; hence, I only call my view *quasi*-error theoretic.) Many have written as if there are objective cognitive standards pertaining to science (arising via objective cognitive aims of science). But I grant that it is open to interpretation exactly which philosophers do, and do not, presume the existence of such things. What matters is that *some* philosophers' claims about first-order goodness makers are false due to their presumptions about cognitive standards. Recall, for instance, the following passage from Kitcher (1993: 92): 'The account that follows will presuppose that there are goals for the project of inquiry that all people share – or ought to share . . . the goals in question are impersonal . . . We need a specification of impersonal goals for science, goals that can ultimately be defended as worthy of universal endorsement.'

If the goals are 'worthy of universal endorsement', then the derivative standards are too. But how could this be unless they were objective? Baptismal intersubjective goals in setting up an institution are plausibly never 'worthy of

---

[47] However, Mackie (1977: 48–9) writes: 'Moral scepticism must . . . take the form of an error theory, admitting that a belief in objective values is built into ordinary moral thought and language, but holding that this ingrained belief is false.' He only says the ingrained belief is false. He does not state that ordinary moral claims are false as a result, although this is the standard interpretation of his position.

universal endorsement', despite the occasional delusions of those responsible for setting them. So to foreground what I argue later, Kitcher seems to be projecting his personal values – albeit values shared by many others in his intellectual circle – into the world. As Mackie (1977: 43) put it: 'We get the notion of something's being objectively good, or having intrinsic value, by reversing the direction of dependence ... by making the desire depend upon the goodness, instead of the goodness on the desire.'

In short, goals are selected because they are valued. Thus, to appeal to goals 'worthy of universal endorsement' is to appeal to values worthy of universal endorsement. But there are no such things, any more than there are Platonic forms. It is a brute fact that we value some things, fail to value others, value some more than others, and so forth. Claims that imply the existence of objective values are therefore false.

Claims implying the existence of privileged intersubjective standards are false, by the same kind of reasoning, if there are no such standards. I will not, therefore, rehearse this point. I will simply argue that there are no such standards in due course.

3.4 On Reactions to the Non-existence of Objective Standards [and Thesis 5]

As Mackie (1977: 34) points out, denying the existence of objective means of evaluation can prompt strong reactions:

> The denial of objective values can carry with it an extreme emotional reaction, a feeling that nothing matters at all, that life has lost its purpose. Of course this does not follow; the lack of objective values is not a good reason for abandoning subjective concern or for ceasing to want anything. But the abandonment of a belief in objective values can cause, at least temporarily, a decay of subjective concern and sense of purpose ... A claim to objectivity has been so strongly associated with their subjective concerns and purposes that the collapse of the former seems to undermine the latter as well.

The idea that there is no scientific progress in any objective sense – or even, as I will shortly argue, in any privileged intersubjective sense – will prompt a similar emotional reaction in some philosophical quarters. For instance, I can imagine someone saying, during an impromptu conversation at a conference, 'Rowbottom is either nuts or insincere. Of course science makes progress independently of what we happen to value, and has progressed a great deal, especially since the Enlightenment. All right-minded first-world intellectuals know this to be true. Science is one of the great success stories of Western democracies: only due to its discoveries do we have nuclear power, spacecraft, and iPhones. Only relativists or charlatans would say that it hasn't progressed!'

Here's how I would respond to such an outburst, were I to overhear it. 'Your underlying concern is that my stance on cognitive progress in science is anti-scientific. More particularly, you think that science is special – in an epistemic sense, among others – and that marking it as progressive is an important way of indicating that. Denying that science is progressive seems, thus, to be putting science on a par with astrology. However, I disagree that I am doing this. I think science is special because it reliably – or most reliably – results in all sorts of cognitive products that pseudo-scientific activities do not. Indeed, there are many *hypothetically rational cognitive aims* of individuals in doing science that there are not, say, in reading tarot. Or to put it differently: *there are many possible desires, in the cognitive realm, that doing science can reliably help one to satisfy (or to bring closer to satisfaction); and many such desires are held by individuals and groups inside and outside science.* One need not invoke "progress" to say this. Moreover, adding "and that's what progress essentially is!" amounts to little more than table thumping. It is relatively empty, just like the ubiquitous contemporary talk of progressive politics. And insisting that progress only concerns something narrower – like achieving changes resulting in greater knowledge – amounts to asserting that one kind of cognitive desire is more important than others. I do not think there is any solid basis for doing this.'

Thus, I grant that talk of science progressing performs a function, in flagging science – especially in the present and the future – as authoritative in an epistemic sense. It encourages people to look to science and scientists to answer certain questions – about whether to be vaccinated against Sars-CoV-2, for instance.[48] This is a good effect. However, the function can be performed just as well – better, with less vagueness and mystery – by pointing out what science has done, can do, and probably will do.

I also see the contemporary realism debate in the philosophy of science as hinging on this reliability issue (and cognate matters), rather than on worries about aims, success, or progress.[49] Perhaps the simplest way to see this is as follows. If someone were to show that science reliably generates approximately true theories about unobservable things, wouldn't that be a great victory for realism? Does something need to be added about aims, success, or progress? Might we not simply note that science is worth doing for those who value said products? The questions are rhetorical.

---

[48] Mackie (1977: 43) similarly thought that in the moral realm: 'there are motives that would support objectification. We need morality to regulate interpersonal relations, to control some of the ways in which people behave towards one another, often in opposition to contrary inclinations. We therefore want our moral judgements to be authoritative for other agents as well as for ourselves: objective validity would give them the authority required'.

[49] For more on this, see Rowbottom (2019: appendix).

3.5 On the Non-existence of Epistemically Privileged Intersubjective Standards [Thesis 2]

Even if there are no objective standards pertaining to cognitive scientific progress, might there be (epistemically) privileged intersubjective ones? Might there be standards selected by an authoritative group (in a broad sense)?

To understand why I answer in the negative, consider again, first, the discussion of aims in Section 2. It is dubious, I argued there, that there are constitutive or essential cognitive goals to the enterprise of science. Moreover, there is no hierarchical structure in science strong enough to support a well-defined group being responsible for the enterprise's goals.

But second, even imagining there is such a group, why do they get to decide what progress *really* consists in? On what basis would their cognitive standards – the ones they happened to impose – trump those of anyone else? Granted, scientific insiders are typically better attuned to what science can reliably do than laypeople, ceteris paribus; hence, they are more likely to select *hypothetically rational aims* in doing science. But beyond that, it is unclear why their standards would be epistemologically privileged.

Perhaps a reader might, at this juncture, think of epistemologists themselves. Why shouldn't they decide on the standards? On the one hand, this is because epistemologists (as a whole) are not well versed enough in science or the history thereof; they are not, that's to say, sufficiently sensitive to how science proceeds. For instance, an epistemologist might be happy to declare that science is all about generating knowledge, on their preferred account of knowledge. But knowledge of what exactly? (Welcome to the realism debate.) The significant worry, here, is that epistemologists are apt to overgeneralise even more than philosophers of science are. And trying to work out something about all inquiry in the abstract – like its aim or what counts as progress therein – is the opposite of the correct philosophical way to proceed. Or so I think. Many other philosophers of science agree, judging by their approaches surveyed in Section 1.

On the other hand, epistemologists do not agree on what constitutes progress. Indeed, the argument for relativity is as strong here as it was concerning objective standards. Dissensus on the nature of first-order goodness makers – and even goodness bearers – is the norm. It follows that even if there is an authoritative group fixing standards: few have identified it *or* few agree on which standards it imposes. Rank incompetence is possibly the cause. A better explanation of the dissensus is that different philosophers impose different standards. To reiterate from Section 3.2, this isn't to say that their standards are *merely* personal; they often derive from the intellectual cultures that their advocates inhabit.

3.6 On True Claims about Cognitive Scientific Progress [Thesis 4]

Not all claims concerning scientific progress presume the existence of objective or intersubjectively privileged standards. And as I alluded to in Section 3.1, talk of progress might be true when the implied standards reflect the values of an individual or a group. Hence, 'Science makes progress when scientific under-standing increases' might be true if the person uttering it takes the reference point as their values (or derivative personal standards). Such a person could disambiguate what they meant by declaring: 'I value scientific changes that increase our understanding.' Evidently such statements can be true, as can statements that instead refer to the values of specific groups.

However, for reasons I explained in Section 2, a rationality requirement is required for the proper deployment of the notion of progress in such contexts. One way of thinking about this is as follows. Valuing an activity for achieving an end is only appropriate when performing the activity connects to – suffi-ciently raises the chance of achieving or furthering – said end. So valuing science for bringing us closer to God, creating world peace, or squaring the circle is inappropriate. The value attributed is misplaced.

In summary, I take second-order cognitive goodness makers to derive from *cognitive values* that are properly applied in so far as they reflect hypothetically rational cognitive aims.

3.7 On Investigative Methods

One of the issues raised in Section 1 was how we should investigate what first-order goodness makers are. I can now provide a brief answer to this, on the view of cognitive progress developed here. I will cover three methods: thought experi-ments, historical studies of science, and social scientific studies of science.

3.7.1 Thought Experiments

First, thought experiments provide a means to explore what we value, and how, if at all, our values are (subjectively or intersubjectively) ranked. For instance, by considering a hypothetical situation where scientists face a dilemma, one might come to see what one cognitively values more. (Neither one's values, nor any hierarchy thereof, are transparent.) It is also possible for a thought experi-ment to encourage someone to alter their values, or their priority ordering thereof, by initiating a process of reflection.

Second, thought experiments provide a means by which to consider what might count as scientifically progressive (given the appropriate reference values), were science able to achieve (or further) that end reliably (or more reliably than other available alternatives). The point here is that science's *actual* ability to do something need not be presumed. Rather, the thought experiments might guide empirical investigations, or alert us to what might, in the future, count as progressive by our own lights.

3.7.2 History of Science

The history of science provides an evidential repository concerning what science can reliably do. For example, theoretical unification or reduction – celestial and terrestrial mechanics becoming one and statistical mechanics accounting for thermodynamics – have occurred repeatedly. Via history we might also, more controversially, discern how much contingency there has been in the way that science has developed.

Historical studies of science can also highlight changes of kinds that we haven't previously considered, by focusing our attention on elements of science that we have not previously paid sufficient attention to. Kuhnian exemplars are an excellent case in point. When one understands the function(s) that these perform – providing shared templates for identifying, formulating, and solving puzzles/problems – it becomes apparent that they might be improved upon (even relative to a fixed set of puzzles/problems). And one might attach cognitive value to such improvements. Finding a faster new way to solve existing puzzles, or a way to solve existing puzzles by deploying fewer intellectual resources, is often considered a boon.

Finally, the history of science is a means of ascertaining what scientists have cognitively valued in the past. For example, the most prominent Cambridge-educated physicists of the nineteenth century valued the invention of non-representational 'models', to foster non-factive understanding about how phenomena interrelate (as discussed in Rowbottom 2019: ch. 4).

3.7.3 Social Scientific Studies of Science

Social scientific studies of science are largely significant for similar reasons as historical studies, but in a different temporal frame. First, they may be used to identify the values of present scientists. Second, they may be used to identify new kinds of change and changes contingently not appreciated via history. Third, they may be used to probe the reliability of processes in, or aspects of, science.

3.8 On Philosophical Progress

I take my findings concerning aims and second-order goodness makers to hold for philosophy too. Indeed, I have provided several tools to explain why philosophers are worried about whether there is progress in philosophy.

Here is my hypothesis. Many philosophers value finding true answers to the tough questions they consider. But they recognise that doing philosophy cannot reliably provide true, or even approximately true, answers to such questions. (Moreover, it doesn't significantly raise the probability of so doing. Getting such answers is not a *hypothetically rational aim* in doing philosophy.) Thus, such philosophers come to worry whether philosophy is progressive after all, by confusing what they (and many of their fellows) happen to value with a standard that *should* be applied to assessing changes in philosophy. The sensible thing to do would be to jettison any talk of progress and just ask what philosophy can and cannot reliably achieve (as against other activities). Then we can work out which of these things we happen to value.[50]

3.9 A Challenging Conclusion

Cognitive progress in science is what one makes it; it is a function of what one values, provided one's values are not misplaced. We have done science because it has provided cognitive progress so construed: progress relative to various sets of standards that different people and groups have had for evaluating changes therein. We continue to do science because it promises more such progress.

Those who think this is wrong – doubtless, there are many – face a challenge. Explain what second-order goodness makers are, if not our values, and specify the appropriate methods for identifying first-order goodness makers as a result.

Those who do not address this challenge but continue to write on cognitive progress could, at least, *roughly* state what they take to be the source of the standards for evaluating progress. This would avert further confusion.

---

[50] It is also possible to start by considering something one values and seeing if it is a hypothetically rational aim in doing something thereafter. But again, this does not need any use of 'progress' or nearby concepts.

References

Aaserud, F. and Heilbron, J. L. 2013. *Love, literature, and the quantum atom: Niels Bohr's 1913 trilogy revisited*. Oxford University Press.

Abbott, B. P. et al. 2016. Observation of gravitational waves from a binary black hole merger. *Physics review letters* 116, 06110.

Bagnoli, C. 2021. Constructivism in metaethics. In E. N. Zalta (ed.), *Stanford encyclopedia of philosophy*. https://plato.stanford.edu/entries/constructivism-metaethics.

Baker, D. 2018. The varieties of normativity. In T. McPherson and D. Plunkett (eds.), *The Routledge handbook of meta-ethics*, 567–81. Routledge.

Bangu, S. 2015. Scientific progress, understanding and unification. In I. D. Toader, G. Sandu & I. Pârvu (eds.), *Romanian studies in philosophy of science*. Springer.

Barnes, E. 1991. Beyond verisimilitude: A linguistically invariant basis for scientific progress. *Synthese* 88, 309–39.

Berker, S. 2019. Mackie was not an error theorist. *Philosophical perspectives* 33, 5–25.

Bird, A. 2007. What is scientific progress? *Noûs* 41, 64–89.

Bird, A. 2008. Scientific progress as accumulation of knowledge: A reply to Rowbottom. *Studies in history and philosophy of science part A* 39, 279–81.

Bird, A. 2022. *Knowing science*. Oxford University Press.

Bohr, N. 1913. On the constitution of atoms and molecules. *Philosophical magazine series 6*, 151, 1–25.

Carter, J. A. and Pritchard, D. 2015. Knowledge-how and epistemic value. *Australasian journal of philosophy* 93, 799–816.

Cevolani, G. and Tambolo, L. 2013. Progress as approximation to the truth: A defence of the verisimilitudinarian approach. *Erkenntnis* 78, 921–35.

Cevolani, G. and Tambolo, L. 2019. Why adding truths is not enough: A reply to Mizrahi on progress as approximation to the truth. *International studies in the philosophy of science* 32, 129–35.

Chakravartty, A. 2017. Scientific realism. In E. N. Zalta (ed.), *Stanford encyclopedia of philosophy*. https://plato.stanford.edu/archives/sum2017/entries/scientific-realism.

Chang, H. 2004. *Inventing temperature: Measurement and scientific progress*. Oxford University Press.

Cohen, L. J. 1980. What has science to do with truth? *Synthese* 45, 489–510.

Cushing, J. T. 1990. Is scientific methodology interestingly atemporal? *British journal for the philosophy of science* 41, 177–94.

Dellsén, F. 2016. Scientific progress: Knowledge versus understanding. *Studies in history and philosophy of science* 56, 72–83.

Dellsén, F. 2018a. Scientific progress: Four accounts. *Philosophy compass.* https://doi.org/10.1111/phc3.12525.

Dellsén, F. 2018b. Scientific progress, understanding and knowledge: Reply to Park. *Journal for general philosophy of science* 49, 451–9.

Dellsén, F. 2021. Understanding scientific progress: The noetic account. *Synthese* 199, 11249–78.

Dellsén, F., Lawler, I., and Norton, J. 2022. Thinking about progress: From science to philosophy. *Noûs* 56: 814–40.

de Regt, H. 2017. *Understanding scientific understanding.* Oxford University Press.

Deutsch, M. 2015. *The myth of the intuitive: Experimental philosophy and philosophical method.* MIT Press.

Douglas, H. 2014. Pure science and the problem of progress. *Studies in the history and philosophy of science* 46, 55–63.

Elgin, C. Z. 2017. *True enough.* MIT Press.

Emmerson, N. 2022. Understanding and scientific progress: Lessons from epistemology. *Synthese* 200: 7.

Galison, P. 1997. *Image and logic: A material culture of microphysics.* University of Chicago Press.

Gillies D. 2000. *Philosophical theories of probability.* Routledge.

Heilbron, J. L. and Kuhn, T. S. 1969. The genesis of the Bohr atom. *Historical studies in the physical sciences* 1, vi–290.

Hetherington, S. 2011. How to know: A practicalist conception of knowledge. Wiley-Blackwell.

Hobbes, T. 1651. *Leviathan.* Andrew Crooke.

Hoyningen-Huene, P. 2013. *Systematicity: The nature of science.* Oxford University Press.

James, W. 1896. *The will to believe: And other popular essays in philosophy.* Longman, Green, & Company.

Keller, E. F. 2002. *Making sense of life: Explaining biological development with models, metaphors, and machines.* Harvard University Press.

Kelp, C. 2014. Two for the knowledge goal of inquiry. *American philosophical quarterly* 51, 227–32.

Kitcher, P. 1990. The division of cognitive labor. *Journal of philosophy* 87, 5–22.

Kitcher, P. 1993. *The advancement of science: Science without legend, objectivity without illusions.* Oxford University Press.

Kitcher, P. 2015. Pragmatism and progress. *Transactions of the Charles S. Peirce society* 51, 475–94.

Kuhn, T. S. 1970. *The structure of scientific revolutions.* 2nd ed. University of Chicago Press.

Kuhn, T. S. 1977. *The essential tension: Selected studies in scientific tradition and change.* University of Chicago Press.

Lagemann, R. T. 1977. New light on old rays: N rays. *American journal of physics* 45, 281–4.

Lakatos, I. 1978. *The methodology of scientific research programmes.* Edited by J. Worrall and G. Currie. Cambridge University Press.

Laudan, L. 1977. *Progress and its problems: Towards a theory of scientific growth.* Routledge.

Laudan, L. 1981. A confutation of convergent realism. *Philosophy of science* 48, 19–49.

Laudan, L. 1984. *Science and values: The aims of science and their role in scientific debate.* University of California Press.

Longino, H. 2001. *The fate of knowledge.* Princeton University Press.

Losee, J. 2004. *Theories of scientific progress: An introduction.* Routledge.

Lutz, M. and Lenman, J. 2018. Moral naturalism. In E. N. Zalta (ed.), *Stanford encyclopedia of philosophy.* https://plato.stanford.edu/archives/spr2021/entries/naturalism-moral.

Mackie, J. L. 1977. *Ethics: Inventing right and wrong.* Pelican Books.

McMullin, E. 1979. Discussion review: Laudan's progress and its problems. *Philosophy of science* 46, 623–44.

Miller, K. and Tuomela, R. 2014. Collective goals analyzed. In S. R. Chant, F. Hindriks, and G. Preyer (eds.), *From individual to collective intentionality: New essays.* Oxford University Press.

Mizrahi, M. 2013. What is scientific progress? Lessons from scientific practice. *Journal for general philosophy of science* 44, 375–90.

Mizrahi, M. 2017. Scientific progress: Why getting closer to the truth is not enough. *International studies in the philosophy of science* 31, 415–19.

Mizrahi, M. 2021. Conceptions of scientific progress in scientific practice: An empirical study. *Synthese* 199, 2375–94.

Mizrahi, M. and Buckwalter, W. 2014. The role of justification in the ordinary concept of scientific progress. *Journal for general philosophy of science* 45, 151–66.

Moberger, V. 2017. Not just errors: A new interpretation of Mackie's error theory. *Journal for the history of analytical philosophy* 5, 1–12.

Moberger, V. 2018. The queerness of objective values: An essay on Mackiean metaethics and the arguments from queerness. PhD Thesis, Uppsala University.

Moxham, N. and Fyfe, A. 2018. The Royal Society and the prehistory of peer review, 1665–1965. *The historical journal* 6, 863–89.

Muldoon, R. 2013. Diversity and the division of cognitive labor. *Philosophy compass* 8, 117–25.

Nado, J. 2014. Philosophical expertise. *Philosophy compass* 9, 631–41.

Niiniluoto, I. 1984. *Is science progressive?* D. Reidel.

Niiniluoto, I. 1999. *Critical scientific realism.* Oxford University Press.

Niiniluoto, I. 2014. Scientific progress as increasing verisimilitude. *Studies in history and philosophy of science* 46, 73–7.

Niiniluoto, I. 2017. Optimistic realism about scientific progress. *Synthese* 194, 3291–3309.

Niiniluoto, I. 2019. Scientific progress. In E. N. Zalta (ed.), *Stanford encyclopedia of philosophy.* https://plato.stanford.edu/archives/win2019/entries/scientific-progress.

Niiniluoto, I. 2020. Social aspects of scientific knowledge. *Synthese* 197, 447–68.

Nye, M. J. 1980. N-rays: An episode in the history and psychology of science. *Historical studies in the physical sciences* 11, 125–56.

Okasha, S. 2011. Theory choice and social choice: Kuhn versus Arrow. *Mind* 120, 83–115.

Oreskes, N. 2019. Systematicity is necessary but not sufficient: On the problem of facsimile science. *Synthese* 196, 881–905.

Orwell, G. 1946. Politics and the English language. *Horizon* 13, 252–65.

Park, S. 2017. Does scientific progress consist in increasing knowledge or understanding? *Journal for general philosophy of science* 48, 569–79.

Popper, K. R. 1963. *Conjectures and refutations: The growth of scientific knowledge.* Routledge.

Popper K. 1983. *Realism and the aim of science.* Hutchinson.

Potochnik, A. 2015. The diverse aims of science. *Studies in history and philosophy of science* 53, 71–80.

Potochnik, A. 2017. *Idealization and the aims of science.* University of Chicago Press.

Psillos, S. 1999. *Scientific realism: How science tracks truth.* Routledge.

Rancourt, B. 2017. Better understanding through falsehood. *Pacific philosophical quarterly* 98, 382–405.

Resnik, D. B. 1993. Do scientific aims justify methodological rules? *Erkenntnis* 38, 223–32.

Rosen G. 1994. What is constructive empiricism? *Philosophical studies* 74, 143–78.

Rousseau, D. L. 1971. 'Polywater' and sweat: Similarities between the infrared spectra. *Science* 171, 3967, 170–2.

Rowbottom, D. P. 2008. N-rays and the semantic view of scientific progress. *Studies in history and philosophy of science* 39, 277–8.

Rowbottom, D. P. 2010. What scientific progress is not: Against Bird's epistemic view. *International studies in the philosophy of science* 24, 241–55.

Rowbottom, D. P. 2011a. Kuhn vs. Popper on criticism and dogmatism in science: A resolution at the group level. *Studies in history and philosophy of science* 42, 117–24.

Rowbottom, D. P. 2011b. Approximations, idealizations, and 'experiments' at the physics-biology interface. *Studies in history and philosophy of biological and biomedical sciences* 42, 145–54.

Rowbottom, D. P. 2014. Aimless science. *Synthese* 191, 1211–21.

Rowbottom, D. P. 2015a. Scientific progress without increasing verisimilitude: In response to Niiniluoto. *Studies in history and philosophy of science* 51, 100–4.

Rowbottom, D. P. 2015b. *Probability*. Polity.

Rowbottom, D. P. 2018a. Scientific realism: What it is, the contemporary debate, and new directions. *Synthese* 196, 451–84.

Rowbottom, D. P. 2018b. Beyond Kuhn: Methodological contextualism and partial paradigms. In M. Mizrahi (ed.), *The Kuhnian image of science: Time for a decisive transformation?* 191–208. Rowman & Littlefield.

Rowbottom, D. P. 2019. *The instrument of science: Scientific anti-realism revitalised*. Routledge.

Rowbottom, D. P. 2021. A methodological argument against scientific realism. *Synthese* 198: 2153–67.

Rowbottom, D. P. 2022. Can meaningless statements be approximately true? On relaxing the semantic component of scientific realism. *Philosophy of science*. 89, 879–88.

Ryle, G. 1949. *The concept of mind*. University of Chicago Press.

Saatsi, J. 2005. On the pessimistic induction and two fallacies. *Philosophy of science* 72, 1088–98.

Saatsi, J. 2019. What is theoretical progress of science? *Synthese* 196, 611–31.

Schweikard, D. P. and Schmid, H-B. 2020. Collective intentionality. In E. N. Zalta (ed.), *Stanford encyclopedia of philosophy*. https://plato.stanford.edu/archives/win2020/entries/collective-intentionality.

Schwitzgebel, E. 2002. A phenomenal, dispositional account of belief. *Noûs* 36, 249–75.

Shafer-Landau, R. 2003. *Moral realism: A defence*. Oxford University Press.

Shan, Y. 2019. A new functional approach to scientific progress. *Philosophy of science* 86, 739–58.

Shan, Y. (ed.) 2022. *New philosophical perspectives on scientific progress*. Routledge.

Stanley, J. and Williamson, T. 2001. Knowing how. *Journal of philosophy* 98, 411–44.

Stich, S. 1993. *The fragmentation of reason*. MIT Press.

Tännsjö, T. 2015. A realist and internalist response to one of Mackie's arguments from queerness. *Philosophical studies* 172, 347–57.

Tuomela, R. 2007. *The philosophy of sociality: The shared point of view*. Oxford University Press.

van Brakel, J. 1993. Polywater and experimental realism. *British journal for the philosophy of science* 44, 775–84.

van Fraassen, B. C. 1980. *The scientific image*. Clarendon Press.

van Fraassen, B. C. 1994. Gideon Rosen on constructive empiricism. *Philosophical studies* 74, 179–92.

van Fraassen, B. C. 1998. The agnostic subtly probabilified. *Analysis* 58, 212–20.

Wallis, C. 2008. Consciousness, context, and know-how. *Synthese* 160, 123–53.

Williamson, T. 2000. *Knowledge and its limits*. Oxford University Press.

Wray, B. 2018. *Resisting scientific realism*. Cambridge University Press.

Acknowledgements

My work on this book was supported by:

* A General Research Fund Grant on 'Scientific Progress: Foundational Issues' (no. 13605620) from Hong Kong's Research Grants Council.
* A stipendiary Visiting Fellowship at Pittsburgh University's Center for the Philosophy of Science.
* A Visiting Fellowship at Cambridge University's HPS Department, funded by a Sino-British Fellowship Trust Grant and a Faculty Research Grant from Lingnan University.

I have benefitted greatly from discussions with colleagues at Cambridge and Pittsburgh, inside and outside seminars. These include Agnes Bolinska, Hasok Chang, Heather Douglas, Stephen John, Edouard Machery, Joe Martin, Matt Parker, and David Wallace.

I have also had useful feedback from audiences at Durham University's CHESS and the University of British Columbia. The Asian Epistemology Network was especially helpful in hosting a seminar which included detailed responses to my talk from Brad Weslake and Xiang Huang.

Several philosophers have also been kind enough to comment on draft sections of the Element – including several that do not appear in the final version! These include Derek Baker, Finnur Dellsén, and Insa Lawler.

I am especially grateful to André Curtis-Trudel, William Peden, and Jan Faye, who worked their way through entire drafts and offered a lot of sensible advice.

Finally, my thanks to Jacob Stegenga for the invitation to write this Element and for his time and patience in helping me to bring the project to fruition.

Any remaining errors are primarily the responsibility of this legion of gifted commentators, who have had ample opportunity to have them corrected. Reviewer one should take the brunt of any blame. Sadly, he/she is anonymous.

Cambridge Elements $\equiv$

Philosophy of Science

Jacob Stegenga
University of Cambridge

Jacob Stegenga is a Professor in the Department of History and Philosophy of Science at the University of Cambridge. He has published widely on fundamental topics in reasoning and rationality and philosophical problems in medicine and biology. Prior to joining Cambridge he taught in the United States and Canada, and he received his PhD from the University of California San Diego.

About the Series

This series of Elements in Philosophy of Science provides an extensive overview of the themes, topics and debates which constitute the philosophy of science. Distinguished specialists provide an up-to-date summary of the results of current research on their topics, as well as offering their own take on those topics and drawing original conclusions.

Cambridge Elements $\equiv$

Philosophy of Science

Printed in the United States
by Baker & Taylor Publisher Services